D0410768

THE LIBRARY OF PRINTING TECHNOLOGY

FINISHING PROCESSES IN PRINTING

Other books in the Series

DESIGN FOR PRINT PRODUCTION
H S Warford

LETTER ASSEMBLY IN PRINTING
D Wooldridge

MATERIALS IN PRINTING PROCESSES
L C Young

MACHINE PRINTING
W R Durrant, C Meacock
R E Whitworth

GRAPHIC REPRODUCTION PROCESSES
Eric Chambers

Series Editor
J E REEVE FOWKES MIOP
Head of the Department of Printing
Southampton College of Art

THE LIBRARY OF PRINTING TECHNOLOGY

Finishing Processes in Printing

BY A G MARTIN

Focal Press, London and New York

ISBN 0 240 50746 0
Printed in Great Britain by
W & J Mackay Limited, Chatham

CONTENTS

Editor's Preface

Printing, in common with other great industries, is caught up in nuclear-age technology. For some printers this is a painful experience: the deep roots of five centuries of tradition resist the pull of technology. The pride of craftsmanship, with all its ingrained mystique, conflicts with the self-assurance of technological skills – skills which reduce the art and mystery of printing to binary numbers.

Already great changes have taken place within the printing industry: more are to come and the rate of change will accelerate. What is happening needs to be understood by printers if they are to remain masters of their destiny. More printers must be trained as technicians and technologists if the benefits of the 'knowledge explosion' are to be fully exploited.

Education and training for the printing industry is undergoing its own revolution to meet the challenge. Courses for printing technicians and technologists have been established: in Great Britain the Institute of Printing has fostered National Certificate, Diploma and Degree courses in printing technology, and the City and Guilds of London Institute offer examinations for a Printing Technician's Certificate.

These new courses need text books, and just as the industry is in danger of being overtaken by events, so too is the printer's library. This was seen by leading educationists, including Charles L Pickering, L T Owens and H M Cartwright. It is largely through their foresight, advice and active help that this new Library of Printing Technology has come to fruition.

The traditional craft of bookbinding has been well served with manuals, but the numerous operations carried out in the so-called 'binderies' of general printing companies throughout the world have been sadly neglected. Indeed, the bindery itself has been the cinderella of the industry – if only because it is last in line in the production sequence.

This book therefore serves a dual purpose: it fills a knowledge gap in the field of print finishing, offering for the production worker a comprehensive textbook on a vast range of processes; for the technician and technologist it explores those areas which are already feeling the impact of the technological revolution.

J E Reeve Fowkes

Acknowledgements

Some of the material in this volume has been culled from other sources and I hope that any individual or manufacturer who recognises material, phrases or sketches that may have appeared in their publications will accept my thanks for allowing it to be used here; a list of publications used will be found in the Bibliography.

E W Pack provided the chapter on box and carton manufacture; A J Brazier read the text and pointed out at least some of the errors that crept in; to these gentlemen, to G W Hales and to other colleagues at the London College of Printing my thanks are due for help and guidance given. But of course all errors and omissions are mine.

During the preparation of this volume reference was made to illustrated literature of the following companies:
Ballinger Rawlings Ltd; Berry, Eade & Co Ltd; Bielomatic (London) Ltd; Book Machinery Co Ltd; British Standards Institution; Borden Ltd (Arabol-Edwardson Division); James Burn & Co Ltd; Camco Machinery Co Ltd; Cameron Machinery Company; Caslon Ltd; R W Crabtree & Son Ltd; Crawley Book Machinery Co; Cross Paper Feeder Co Ltd; Cundall Folding Machine Co Ltd; The Collmar Corporation; Dispro Ltd; Fasson Products Ltd; Oscar Friedhelm Ltd; Fryer Edwards Ltd; Funditor Ltd; Geliot Hurner & Ewen Ltd; General Binding Co Ltd; Gildmore Press Ltd; Graphic Arts Equipment Ltd; Hare & Co Ltd; Hunkeler Ltd; Letraset Ltd; J Marshall Ltd; Miehle Goss Dexter Ltd; Minnesota Mining & Manufacturing Co Ltd; Morane Plastics Co Ltd; Muller Martini Ltd; National Adhesives & Resins Ltd; PIRA; Planax Binding Systems Ltd; Price Service & Co Ltd; Radyne Ltd; Rollem Patent Products Ltd; Schuler Sales & Service Ltd; Harris Intertype Ltd (Sheridan Division); Smyth Horne Ltd; Stanelco-Thermatron Ltd; Soag Machinery Co Ltd; Tragacine Adhesives Ltd; Triumph Radio Frequency Ltd; Vacuumatic Ltd; Vesco Ltd; Vickets (Engineering) Ltd; Victory Kidder Co Ltd; G M Whiley Ltd; Worsley-Brehmer Ltd.

In dedicating this book to my wife, I have in mind her tolerance whilst I spent long hours at the typewriter and drawing board.

A G Martin

Introduction

The term 'print finishing processes' is of fairly recent origin and has been coined to provide a generic title for the wide range of pre-printing and post-printing operations that transform paper and raw printed sheets into a saleable article. At the least this may mean no more than reducing large printed sheets to a smaller size by a cutting operation; while, at the other end of the scale up to twenty separate operations may be applied to produce an elaborate and expensive 'binding'.

Historically much of this work was concerned with the bookbinding process and for this reason some finishing departments still have the term 'bindery' applied to them; although that department may not have completed any true binding work for many years. In some printing houses the completion work was carried out in the 'warehouse' along with storing of paper and despatching operations and, although very undescriptive of the type of work executed in this section, 'productive warehouse' as a term has been with us for some considerable time.

There is, of course, a great deal of specialisation in company activities and this allows the use of highly productive machinery. Even so, some printing managements expect the finishing section to cope with a wide range of products and this frequently means the installation of many small machines that may only be used for quite short periods of time. Three main areas of production can be identified as print finishing, binding, and box and carton manufacture. Typical products in each are:

1 Print finishing

Storage and preparation of white paper; cutting to size both white and printed stock; production of booklets and magazines, stationery and file material, labels, showcards and calendars; assembly work, *eg* boxing and set-making; gumming, varnishing and laminating; packing and despatch.

2 Bookbinding

Letterpress, *eg* printed books such as, paperbacks, publishers' case work, bibles, pocket and other diaries; library rebinds and other repair work; fine leather work. Stationery, *eg* writing books such as ruling for exercise books; flush work; ledgers; loose leaf binders; plastic binding devices and mechanical styles; swatchbooks; and miscellaneous services including blocking, box and file manufacturer, etc.

3 Box and carton manufacture

Boxes, *eg* rigid containers; drums and tubes; stitched, covered and presentation boxes. Cartons; folding cartons of all types. Cutting and creasing; showcards and display work; jigsaws; and calendars.

There is considerable overlap between (1) and (2) and between (1) and (3). Some simple forms of binding, *eg* paperbacks, school primers and magazine work, may be produced in either the bindery or print finishing section; similarly jigsaw puzzles, cut shapes for display work and cutting and creasing may be found in both finishing and carton-producing departments.

Within each major division specialist subdivisions exist: many 'carton' houses may have nothing to do with 'boxes'. In another field some companies specialise in the production of publishers' case bindings whilst others simply fold, envelope and mail other people's printing. By this rational application of machinery the large capital investment may be recovered by a high utilisation rate.

There is, of course, a considerable market for paper products having little or no printing involved in them; these are generically referred to as 'converting'. The stage at which conversion becomes 'finishing' puts a very fine point on the problem and some companies are in both fields. Converting includes the making of envelopes, pads and other stationery articles; tissue conversion; tubes and other cylindrical shapes; cups and other paper table stationery; sack and bag manufacture, etc. Any printing that may be involved in these products is, in general, likely to be an inherent part of a continuous production process and of a simple nature.

As a general rule most of the skilled machine work in finishing operations is carried out by male labour with semi-skilled male or female assistants where these are needed. There are more female workers in finishing than in any other area of the industry, largely because of the many simple repetitive tasks that have to be performed and at which the ladies excel. Nevertheless, many traditionally female tasks are highly skilled and particularly suit the delicate dexterity that female workers can offer – for example, hand book folding, sewing, sheet mending, etc.

There is an increasing tendency on the part of manufacturers to integrate processes that were formerly separate hand operations, with the consequent displacement of unskilled labour. An example of this can be seen in the production of wire-stitched booklets which may originally have been inset and stitched as separate operations, finally being trimmed on a single-knife guillotine. Today the job may go on to an in-line machine that will inset, stitch and trim at high speeds and need a minimal number of unskilled assistants for loading and taking off.

Part I of this book is an introduction and aims to cover in a general way most of the processes and principles underlying print finishing. Part II looks at more specialised aspects and particularly those areas embodying the use of technological principles as distinct from the application of craft skills.

1. Paper management

Paper is the prime material in printing and constitutes a large part of the cost of production. It is therefore very necessary that paper stocks are managed in an efficient way.

Paper merchants will normally supply stock lines wrapped in suitably labelled parcels containing 250, 500 or 1000 sheets according to size and weight. Larger quantities may be delivered in suitable stacks, strapped to pallets ready for off-loading mechanically.

THE PAPER STORE

The paper store should be strategically sited in the building so that receipt and subsequent delivery to the machine room is unhindered. Unless it is to be used almost immediately paper is racked, labels showing and suitably flagged, in, say, units of ten reams, for ease of counting.

Racks of metal 'L' section, tubular section and of wooden construction are used; the use of slotted angle allows the racking to be reformed to meet the changing needs of the company. A careful survey of the company's requirements will reveal the normal stock sizes most used and the racks are constructed accordingly. High racks are possible for light parcels of small area but solid shelving is necessary to keep the paper flat and to allow the air to circulate. The bottom rack should be some 100 mm off the floor for similar reasons. Space between racks should be sufficient to allow suitable mechanical handling equipment to be operated.

Furnishing of the store will include sufficient portable benches and ladders to prevent these having to be moved excessively. Sometimes mobile racking is suitable for the paper store and this may considerably increase the amount of paper held in a given area. In this case the racking is mounted closely in banks, riding on rails or overhead slides. Access to the racks in the second and subsequent banks is by sliding the front rack sideways and in some installations this is done in sections.

The stock system used will reflect the size of the installation and may be either a simple stock book, a loose leaf wall-mounted or free-standing card index system. The design of the system will enable stock to be entered on receipt, written off against job numbers and a running total maintained. Other information entered will include description, manufacturer, supplier of the item and its location in the store. Large installations sometimes duplicate the system in the general office so that administrative personnel may have access to the stock state.

All paper tends to deteriorate while in stock and some may become unusable in quite a short time. Minimum stocks should be maintained and stock should be 'turned over' as frequently as possible. Paper stock should be the whole responsibility of one individual and only issued on receipt of suitable authority.

Atmospheric control in the paper store

The atmospheric environment in which paper is stored may have a fundamental effect upon the shape and condition in which it reaches the printer. Ideally the conditions should be rigidly controlled, but this is seldom possible due to the high cost of adding the necessary equipment to an existing building. Even in new building ventures this may be seen as an expensive luxury.

While racking the paper in mill wrappings may protect it from some of the vagaries of the atmosphere, wide changes in relative humidity will result in stresses in the edges of the stacks giving either waving and curling edges or tight edges and loose centres. Both these conditions will give rise to difficulties on the printing machine.

Opinions vary on what is considered to be the ideal atmospheric condition, but standards between 55–65 RH at temperatures between 20 and 21°C are usual.

These temperatures are similar to those required in the factory workshops and it is usually possible to link-in some form of thermostat to maintain these levels during normal working periods. But of course a working week of 40 hours represents only 25 per cent of the weekly total and if the factory heating is off during the remainder of the time it may be necessary to ensure that the paper store has some form of subsidiary heating to avoid extremes of temperature.

Partial control of relative humidity can be achieved by the installation of humidifying equipment, some of which is portable and simply plugs into the mains electrical supply. Larger units are fixtures and usually require both water and heating services. A humidistat measures the humidity in the store and if the moisture content is too low it instructs the unit to pump out vaporised water until the correct standards are reached.

Even without sophisticated equipment the careful stock-keeper may be able to minimise excessive variations by manual control of heating and ventilation. In this he will be assisted if suitable thermometers and hygrometers are sited in the store.

Paper conditioning

While environmental control may be highly desirable in the paper store its presence does not necessarily preclude the conditioning of paper after passing out of the store and before being printed. Work which requires the paper to be stabilised to ensure accuracy of register is frequently conditioned in the machine room itself. One method uses a conveyor chain moving very slowly 4–5ft above floor level and this has ball clips into which a number of sheets are pushed. This system allows the paper to hang in the printing-room atmosphere for up to two hours while the conveyor is slowly moving forward. It is then removed from the chain and is ready to be loaded on to the press.

Handling

Paper is a very heavy material and mechanical handling methods should be used wherever possible. Large quantities that have to be moved considerable distances probably warrant movement by fork-lift truck, but of course the factory conditions must be suitable. These very mobile pieces of equipment are available in an extensive range of sizes and types; they will increase the height of stacking but require much wider gangways for manoeuvring than is usual in the paper store.

Jacking trolleys or pallet trucks are more familiar in the smaller and medium-sized companies. These lift the loaded pallet off the floor a few inches by mechanical or hydraulic means and then the load is manually pushed to its new location.

Pallets are the flat portable platforms on which goods are transported. Various pallets are available and are specially designed for use with fork trucks and other pallet-handling appliances both for storage and through transportation (fig. 1.1). Stillages (or skids) have a top deck similar to the pallet but are supported by steel or wooden legs (fig. 1.2).

A wide variety of hand-pushed wheeled trucks, trolleys and boxes are used to transport paper from point to point and it is sometimes convenient to start the job from the paper store in such a conveyance, staying with it until completion.

Printed stock store

Many companies operate a department with this title and a common feature

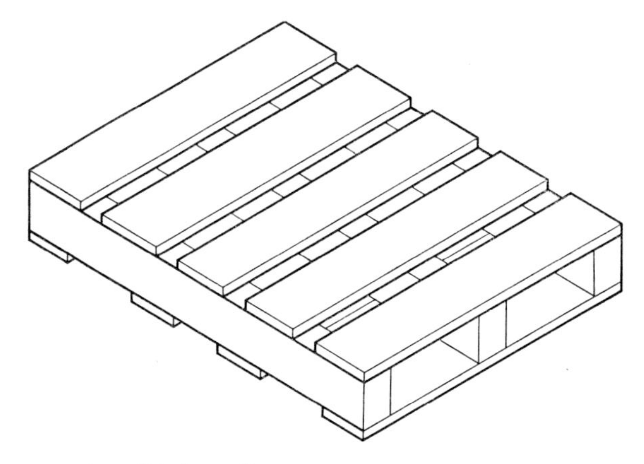

1.1 *Two way opening, all timber pallet.*

is that the material held in the store is often the legal property of the customer, the job having been paid for and delivered into the printed stock store instead of into the customers warehouse.

In printing houses the store will contain work that requires overprinting or updating, *eg* labels, posters, showcards and booklets held awaiting the customer's further instructions. Binderies often hold considerable stocks of bound and unbound books on behalf of the publisher. Work that has been printed or otherwise processed elsewhere (outwork) may also be held in this

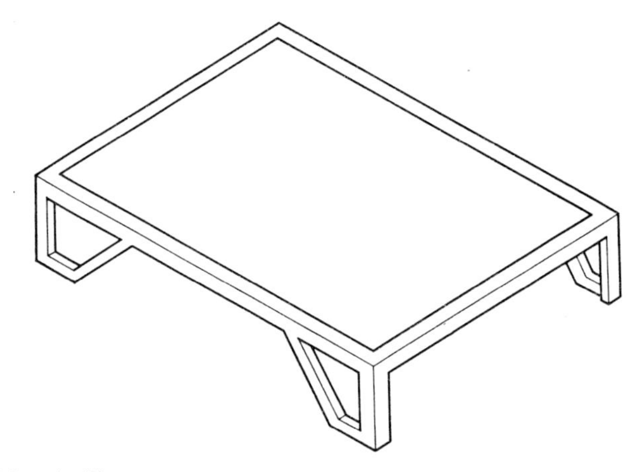

1.2 *Steel-legged stillage.*

section while waiting to be fed into the mainstream of production in the factory.

As this is a service provided for the customer, and sometimes charged for, it is important that correct and accurate stocking procedures are maintained. Particularly important is the preservation of printed stock in fair condition and steps must be taken to prevent deterioration by excessive light, dust and vermin.

PREPARATION FOR PRINTING

The task of the white-paper warehouseman is to receive stock, take care of it while in his paper store and to issue for each job the quantity required to correct size. The latter task may involve both counting out the required number of sheets and then cutting them to job size.

Counting devices are of recent origin and even today some short-run counting may be quicker by hand. Where the application is suitable, the

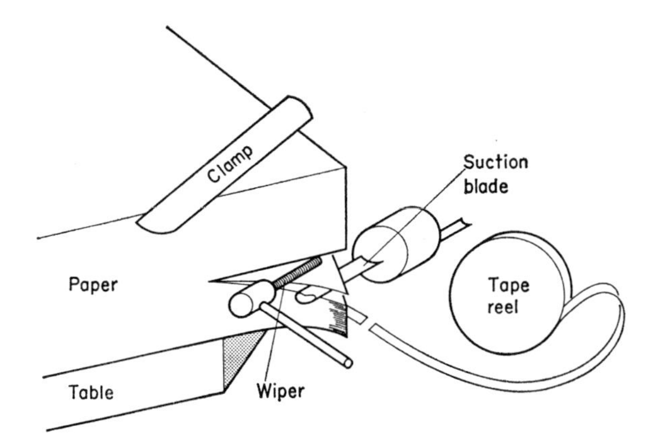

1.3 *Paper counting head with electronically controlled tape insert. The corner of the paper stack to be counted is presented to the counting head so that the paper covers the hole in the suction blade. When the lowest corner has been pulled down by the suction the wiper arm moves elliptically above the blade and tucks the corner downward simultaneously the machine records one unit. Immediately the second corner becomes accessible to the suction and is similarly treated. The electronic counter is previously set for batch number and when this number is reached the tape insert mechanism is actuated. The end of a narrow reel of tape is cut off and inserted at high speed into the stack. The counting head continues to rise in front of the stack as necessary, counting and inserting the paper tabs. On counting the last (top) sheet the vacuum fails and the mechanisms come to a halt. The final count is recorded as a number of batches and residual units.*

counting device may be either bench or floor mounted; they are usually electro-mechanical in character and operate on the corner of the pile. The control unit may be set to count and tab in suitable numbers and the final score can be read off on the control panel. An accuracy of 1 or 2 sheets per thousand is obtainable and this is usually good enough for most commercial applications. For counting 'security printing', machines with a guaranteed 100 per cent accuracy are available (fig. 1.3).

Paper sizes and subdivisions

The application of the metric system in the United Kingdom means the adoption of the ISO paper sizes by the paper and printing industries. Although it is intended to phase out the traditional British paper sizes fairly rapidly it

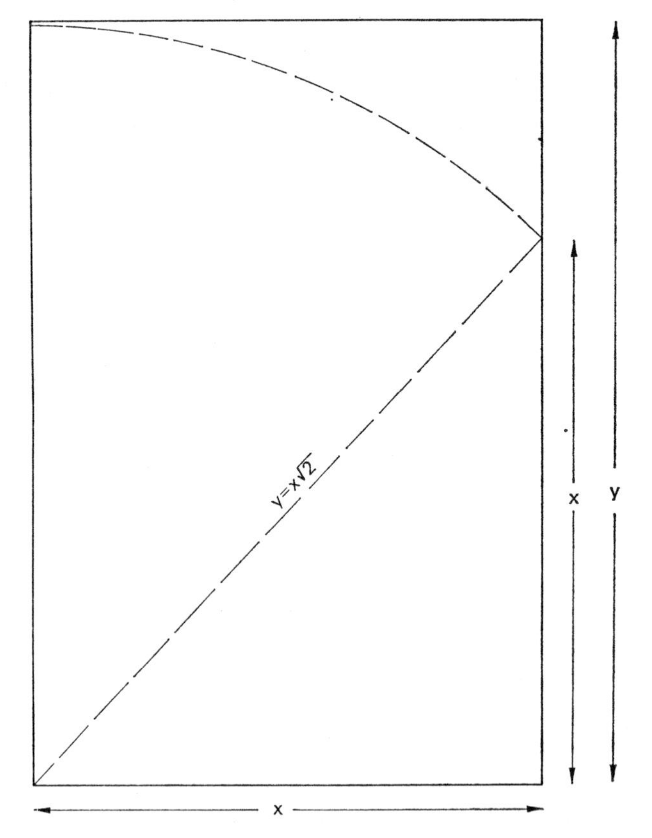

1.4 *Proportions of ISO sizes.*

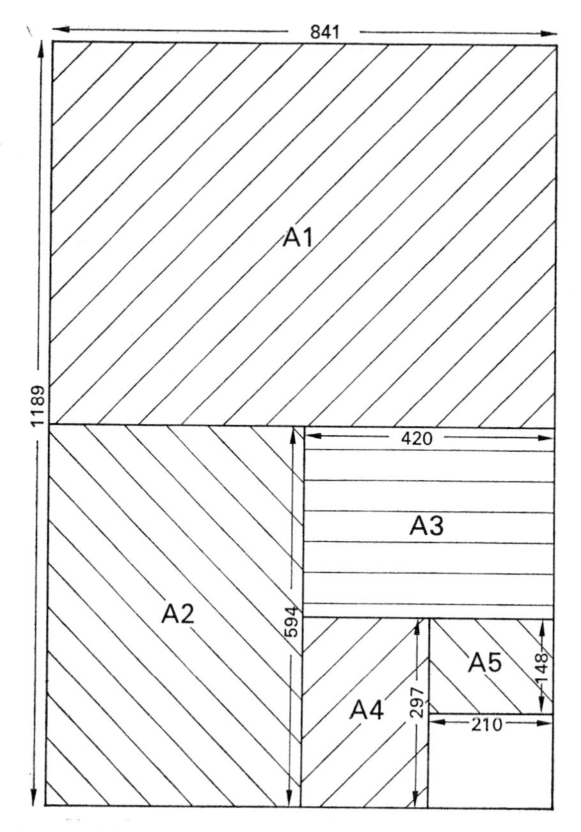

1.5 *Regular divisions of Ao.*

is probable that one size, crown and its derivatives, will be with us for some time.

In the ISO range three basic series of sizes are used; these are trimmed (or finished) sizes and are designated 'A' ,'B' and 'C'. The first two have a common proportion basis which is expressed as $1:\sqrt{2}$ or $1:1{\cdot}414$. With these proportions the long-edge dimension is the length of a diagonal of a square formed by the smaller dimension of the sheet (fig. 1.4), and dividing the sheet progressively by two preserves the proportions.

The A series is based on the dimension 841mm × 1 189mm (Ao) which is one square metre in area. Two larger sizes are calculated by doubling the short dimension of Ao to give 2Ao and by doubling both dimensions resulting in 4Ao.

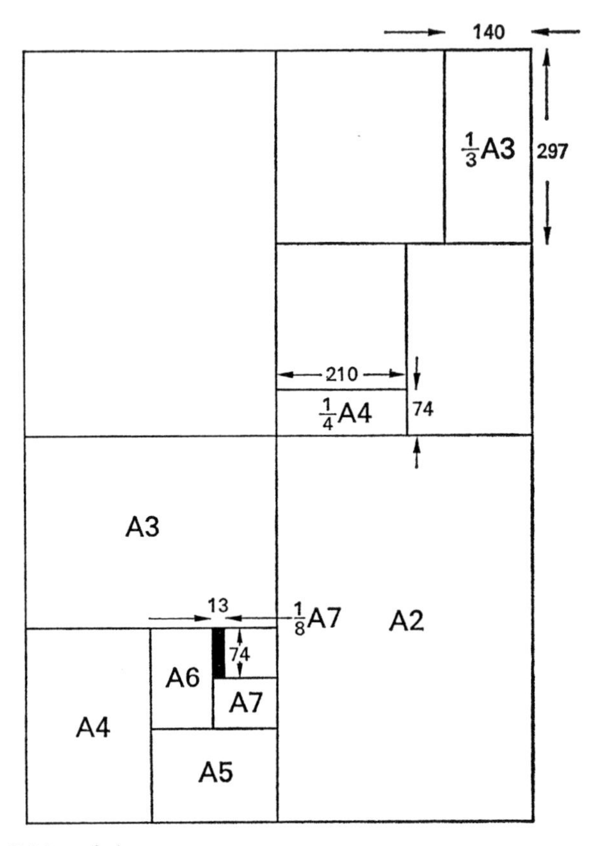

1.6 *Irregular divisions of Ao.*

Subdivisions of Ao are designated A1, A2, A3, etc, the figures denoting the number of times the basic Ao sheet has been divided into two parts. Figure 1.5 illustrates the division of Ao down to A5 (148 mm × 210 mm).

Long divisions are calculated by further dividing the A subdivisions parallel to the short dimension (fig. 1.6), *eg*

$\frac{1}{3}$ A3 = 140 × 297 mm
$\frac{1}{4}$ A4 = 74 × 210 mm
$\frac{1}{8}$ A7 = 13 × 74 mm

The 'B' series is recommended for 'posters, wall charts and similar items where the difference in size of the larger sheets in the A series represents too large a jump' (BFMP *Going Metric with the Printing Industry*) but 'it is felt that

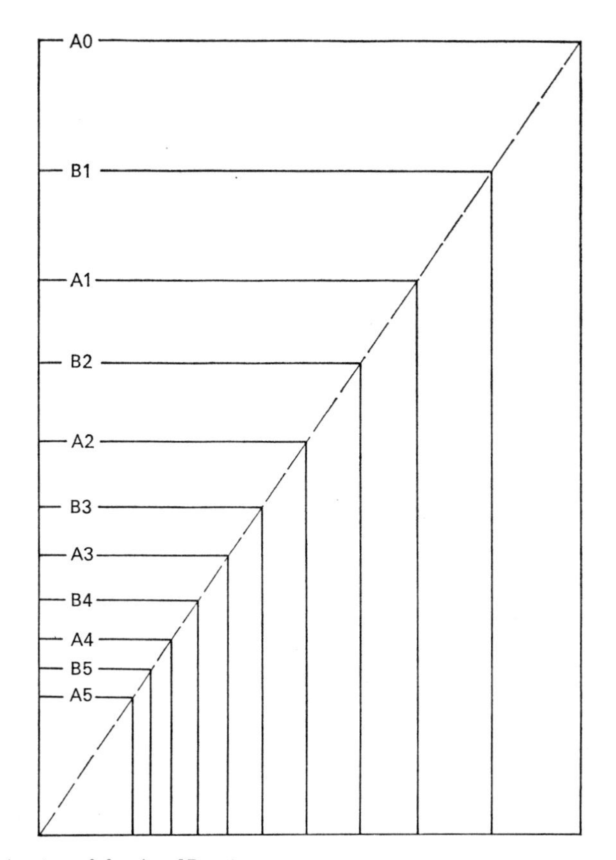

1.7 *Relative sizes of the A and B series.*

these sizes should not come into general use in the United Kingdom' (page 6, BS 4000:1968). This series of sizes is obtained by placing the geometric mean between the adjacent sizes of the A series in sequence so that the same proportions are obtained (fig. 1.7). The basic sheet size B0 is 1000mm × 1414mm; subdivisions and long divisions are obtained as in the previous series.

The 'C' series is for envelopes or folders and the sizes are designed to accommodate the A subdivisions, *eg* a C4 folder will enclose A4 paper.

It should be noted that the following scale of tolerances is laid down in BS 4000:

For dimensions up to and including 150mm	± 1·5mm
For dimensions greater than 150mm and up to 600mm	± 2mm
For dimensions greater than 600mm	± 3mm

The white-paper storekeeper will be required to stock a range of paper selected from four basic series; the trimmed A and B series already mentioned and the larger 'RA' and 'SRA' series.

For normal trimmed work paper from the RA sizes would be used, *eg* if RA0 is printed 32 pages up, folded and trimmed to A4, the fore-edge trim will be 5mm, the head trim 4mm and the tail trim 4mm.

Larger trims and grip are provided on the SRA series and in the same situation as above the trims would be fore-edge 15mm, head 11mm and tail 12mm. The SRA series is especially useful for bled work, *eg* where illustrations are imposed to be cut into at the finishing stage and in this case the blocks should not be less than 2mm larger than the trimmed size of the job. Accepted tolerances for the RA and SRA series are between ± 3mm and ± 5mm according to size.

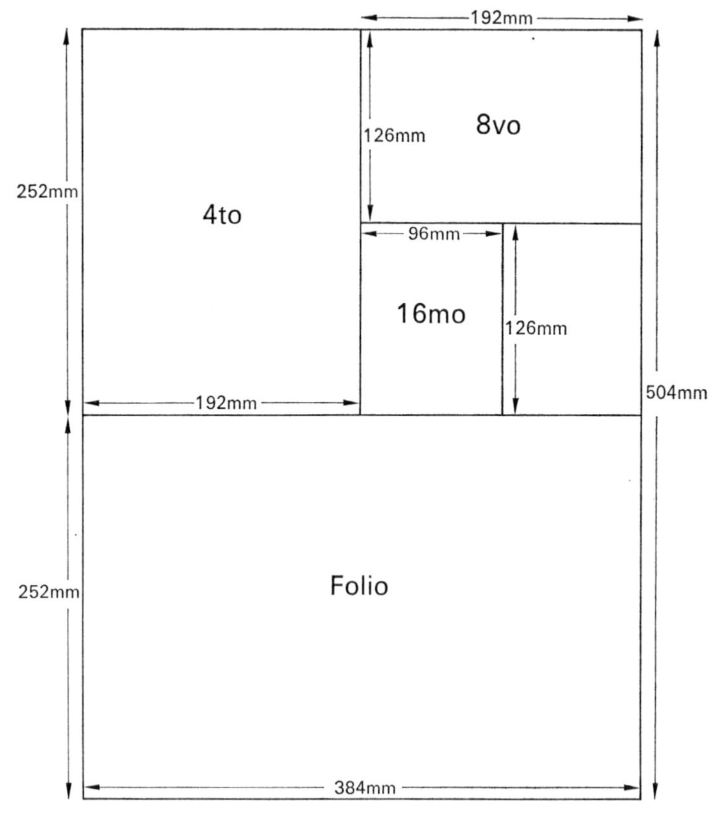

1.8 *Subdivisions of Metric Crown paper.*

The principal sizes likely to be stocked by the paper merchant are A0, A1, A2, RA0, RA1, RA2, SRA0, SRA1, SRA2 and the paper store will hold a selection of these to suit the nature of the work normally completed by the company.

Agreement has been reached to offer a limited range of substances extracted from the R20 series of preferred numbers. These will probably be 28, 31·5, 45, 63, 71, 100, 140, 180 and 200 gm² plus 85, and 160 gm² from the R40 range and 25, 50 and 160 gm². The last three are being added specifically for lightweight papers, continuous stationery and systems cards respectively.

With nine basic sizes, up to fourteen substances and innumerable types of paper quality, the permutations are extremely wide and the total number of different stocks held in the paper store may be no less than was the case with the British paper sizes.

Metric book sizes

The ISO paper sizes which are internationally recommended for general printing are not very suitable for books and are rarely used. New recommendations have been formulated to meet the special requirements of the book-manufacturing industry and these take cognisance of the existing conditions. The main nomenclatures are retained recognising that a series of sizes in octavo is necessary and desirable.

Three basic British Standard paper size names are used plus one well-known non-standard name. The original measurements of these were as follows:

	Single	*Double*	*Quadruple*
Crown	508 × 381mm	762 × 508mm	1016 × 762mm
Large crown*	533 × 406mm	812 × 533mm	1066 × 812mm
Demy	572 × 445mm	889 × 572mm	1143 × 889mm
Royal	635 × 508mm	1016 × 635mm	1270 × 1016mm

*Non-standard.

Subdivisions are obtained by folding or cutting into

2 parts = folio (fo)
4 parts = quarto (4to)
8 parts = octavo (8vo)
16 parts = sextodecimo (16mo) (see fig. 1.8)

British Standard sizes untrimmed were

crown folio = 381 × 254mm
crown quarto = 254 × 190mm
crown octavo = 190 × 127mm
crown sextodecimo = 127 × 95mm

In the same way subdivisional dimensions were calculated for the other standard sizes.

The new recommendations rationalise the quadruple sizes with each metric dimension divisible by both 8 and by 4; this allows imposition to be accomplished without the need for fractions of a millimetre to be used. Quad crown has been adjusted from 1016 × 762mm to 1008 × 768mm and this size is referred to as 'metric quad crown'. Similar adjustments are made to large crown, demy and royal. The five recommended standard metric book sizes are:

	*Trimmed size**	*Untrimmed size*	*Quad paper sizes*
Metric crown 8vo	186 × 123mm	192 × 126mm	768 × 1008mm
Metric large crown 8vo	198 × 129mm	204 × 132mm	816 × 1056mm
Metric demy 8vo	216 × 138mm	222 × 141mm	888 × 1128mm
Metric royal 8vo	234 × 156mm	240 × 159mm	960 × 1272mm
A5	210 × 148mm	215 × 152·5mm	860 × 1220mm (RA0)

*Standard trim is 3mm except on the A5 size

From the above table other metric subdivisions may be calculated, *eg* metric crown quarto is double the size of untrimmed octavo short dimension = 252 × 192mm untrimmed or 246 × 189mm trimmed.

In each case the new metric octavo trimmed sizes are within a few millimetres of the old trimmed sizes and little difference will be noted by the book-buying public. The production of non-standard sizes is not affected in any way.

When writing down paper sizes on order forms it has been the practice to indicate the binding or left-hand edge of the paper as the first dimension, *eg* 186 × 123mm is a broad or upright shape and when written 123 × 186mm a landscape or oblong shape is intended (fig. 1.9). The ISO symbols, *eg* A5, imply a portrait shape. Methods of indicating landscape shapes are to be given in BS 2489 which is at present being revised.

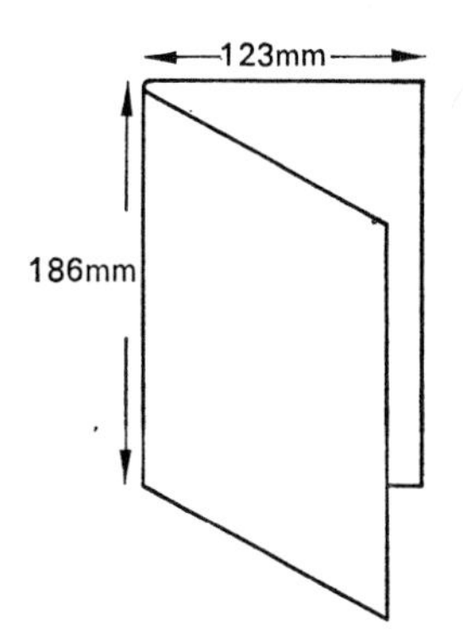

1.9.1 *Broad or upright shape.*

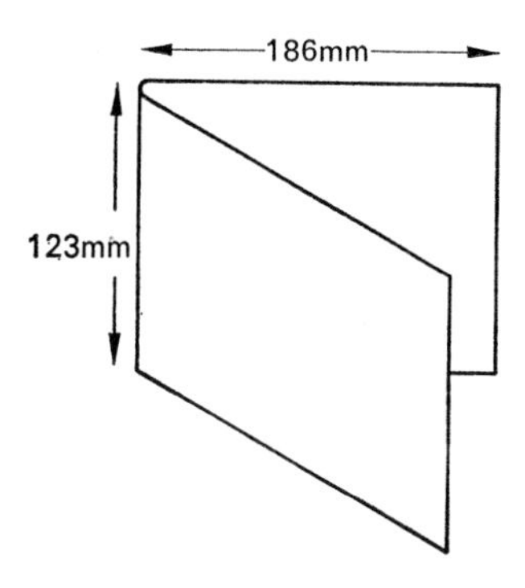

1.9.2 *Oblong or landscape shape.*

Cutting

When cutting out subdivisions from full-sized sheets consideration is given to the following factors.

1 Is the quoted size 'finished', or is the job to be further trimmed after printing?
2 Are clean square edges necessary for printing purposes, *eg* the register of colour work, work and turn techniques, etc?
3 Is the way of the 'watermark' or 'grain' important?
4 Is the job being printed more than one up?

Although the computation of cut sizes should be the task of the order office it is often desirable for these to be checked and within these limits the skill of the guillotine operator will determine the accuracy of the paper dimensions sent forward for the printing process. On a manually operated machine, very high degrees of consistent accuracy are difficult to obtain, particularly in straight splitting of paper, *eg* dividing a ream into say eight parts. A small error in setting or reading the tape (parallax error) will be doubled in the visual difference of the two halves of the job. Although modern machines with micrometer settings produce greater accuracy, 'double dressing' is often used when identically sized sheets is important. This implies that after splitting into oversize subdivisions, the edges produced by the bevel side of the knife are put into the machine and trimmed again to the finished size.

'Squareness' of trimmed paper is obtained by utilising the side plates of the machine which are known to be at 90 degrees with the cutting line.

Guillotines

The range of guillotines available to the print finisher is probably greater than any other single machine with the exception of folders. The cheapest machines are those having hand clamp and hand levered cutting actions with openings of around 500mm. Powered machines also start with relatively narrow cutting openings, continuing through a range which terminates in the large powerful machines used by the papermills for trimming their largest sheets.

Typical machines for use in conjunction with the paper store are those with cutting openings of 1070mm and 1370mm, but the machine should, in any case, be 100mm wider than the diagonal dimension of the largest sheet to be cut.

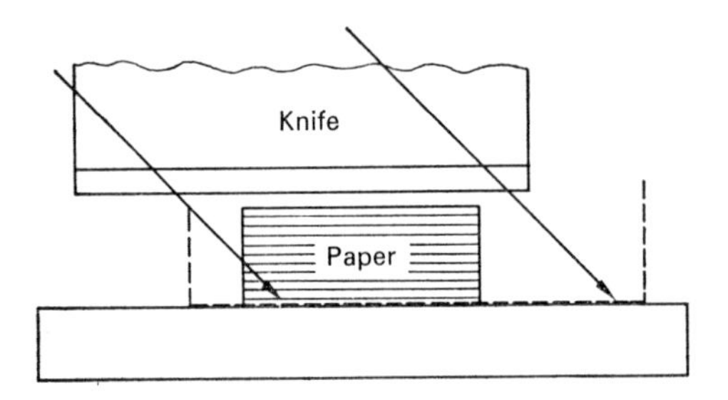

1.10.1 *Oblique cutting action.*

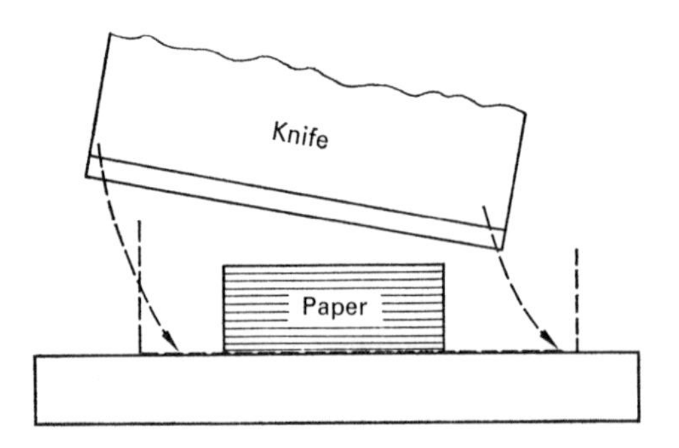

1.10.2 *Dip shear action.*

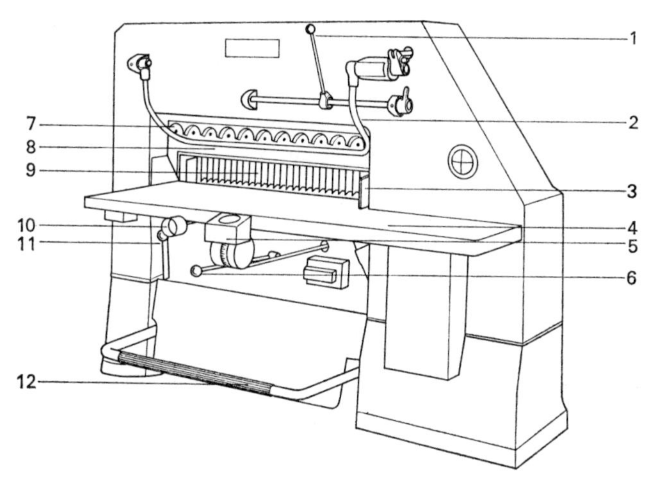

1.11 *Modern standard guillotine.*

1 *Starting handle*
2 *Sweep away guard*
3 *Side plates*
4 *Bed or table*
5 *Magnifying tape reader*
6 *Back gauge control handle*
7 *Knife beam*
8 *Knife*
9 *Back gauge*
10 *Fine adjustment control*
11 *Backgauge locking lever*
12 *Clamp foot-control pedal*

Machines that embody a vertically acting knife are now seldom seen and the modern guillotine will utilise an oblique or a dip-shear action (fig. 1.10) and manufacturers using the latter claim an economy of power used and a more satisfactory cut. Most modern machines embody some or all of the following characteristics (fig. 1.11):

1 The knife fall is so engineered that the cut is made at 90° with the table. Adjustments to take up wear should be available.
2 Clamping should be adequate and adjustable to various levels to suit different types of stock. The clamp action should be complete before the knife strikes the paper.
3 The back gauge must be easily moved by hand or be powered. It should be easily locked in position and when locked immovable. The back

gauge should be parallel to the cutting line and adjustable both vertically and horizontally. For very small cutting distances interlocking with the clamp is desirable.

4 Knives should be of good quality and of good depth to allow many regrinds: easy and swift knife change facilities should be built in, with fine setting adjustments.
5 Cutting sticks should be easily adjusted, changed or turned; but once set, immovable.
6 Side plates should allow the cutting of small areas accurately.
7 There should be accurate gauging and presentation of the cutting distance in a position which is easily read.
8 Suitable safety guards should be fitted which conform to the Factories Act and Regulations. Powered machines often incorporate some device that will slip or break if the machine becomes accidently overloaded. This should be easily reset.

Other facilities offered on modern guillotines include micrometer settings, air cushion tables, illuminated tables, semi-automatic and automatic methods of controlling the back gauge, and photo-electric cell operated guards.

A key factor in good guillotine practice is the knife and two of these are usually provided with each machine. As the knife and its mount have to absorb the initial shock when striking the paper, the metal from which the knife is constructed requires to be of a tough character, with a low-grade steel or iron forming its bulk. Inset into the critical part of the knife is a piece of high-grade steel and it is this which actually cuts and which is resharpened (fig. 1.12).

1.12 *Guillotine knife with inset.*

Knives are also made from single units of medium carbon steel and from chrome and other alloy steels. Where it is appropriate, heat treatment is applied to produce the correct degree of hardness and toughness.

In general terms it can be said that knives should be bevelled to suit the type of work being cut. Guillotines often have to cut a wide range of work, so a compromise bevel is used. A bevel of 16° may be suitable for soft paper stock, 19° for general work and 22° for harder materials. Bevels of less than 16° leave the edge weak and liable to chip; more than 22°, although very robust and long lasting, places an intolerable strain on the mechanisms.

Bevels may be flat or hollow ground, the former being preferred for strength and lasting quality. Double bevels are sometimes specified and are particularly useful when exceptionally hard stocks are to be cut (fig. 1.13).

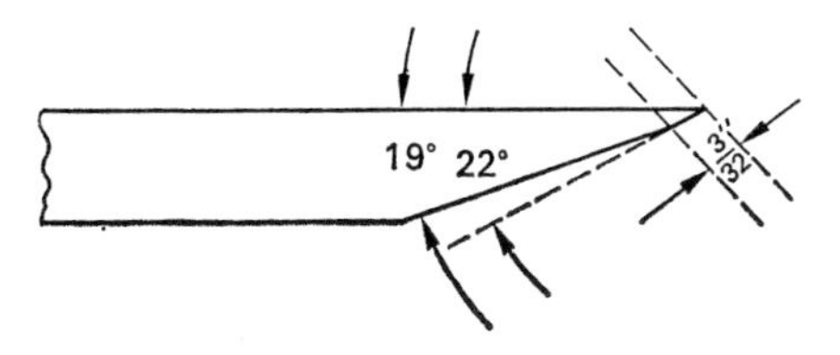

1.13 *Double-bevelled edge.*

Honing of newly ground and partly used knives is widely practised, a very fine grade flat oilstone being used. It is important that the non-bevel side should remain flat and the bevel evenly honed to prevent non-cutting hollows appearing. Special honing devices are available that carry the stone at a predetermined but adjustable angle and these allow the knife to be honed *in situ*.

Very blunt knives place unnecessary wear on the machine and indications of this condition are excessive impact noise, blocking or sticking of the cut sheets, heavy burr on the trimmed edges, small creases on the face of the pile where the cut starts and cracking noise when the last sheet in the pile is sheared.

Quality control

Before leaving the paper store the paper that has been prepared for a particular job will be checked for quantity, type and substance, colour and dimension. A suitable document will be attached to the stack showing the job number and other relevant details before it is placed in the 'paper ready' area of the store.

2. Folding

A sheet of printed paper will often require to be folded or cut to reduce it to the size ordered by the customer. Leaflets and advertising brochures may need only one or two folds, a book or magazine three folds and maps as many as six, seven or more.

A printed sheet of paper will have the following characteristics that have an effect upon the folding:

1 The forme should be centred on the sheet, unless there is good reason for it not to be so.
2 Back-up should be accurate, the back mark in correct position and the signature properly located.
3 The size of the sheet should remain constant throughout the job and if the sheet has been slit on the machine the resulting edge should be clean and accurate.
4 Grip and lay should be clearly marked.

Some work will be 'folded to paper', particularly where the area of print is irregular and the pages do not superimpose upon each other in any way. Greetings cards and book endpapers are examples of work that is folded this way.

Books and magazines are 'folded to print', great care being taken to ensure that type areas fall one above the other accurately.

Hand folding

Cost of labour makes all hand work expensive and this is true of folding too. It is usually a female task and is mostly confined to short runs, odds, impositions for which the company have no machine and very expensive printing where spoilage must be negligible.

The simple tool is the folding bone, a flat piece of polished bone 250mm long with rounded ends. A 16-page section requires three right-angle folds and these are made with the paper remaining flat on the work table. Producing right-angle work beyond two folds on medium and thick paper results in an unsightly crease appearing at the head of the centre of the section (fig. 2.1). This is caused by the inability of the centre pages in the section to move forward to accommodate their own thickness and this results in a bunching and creasing in the paper. When hand folding this fault may be eradicated by splitting the head bolt, rather more than halfway along the sheet with the bone folder before completing the last fold.

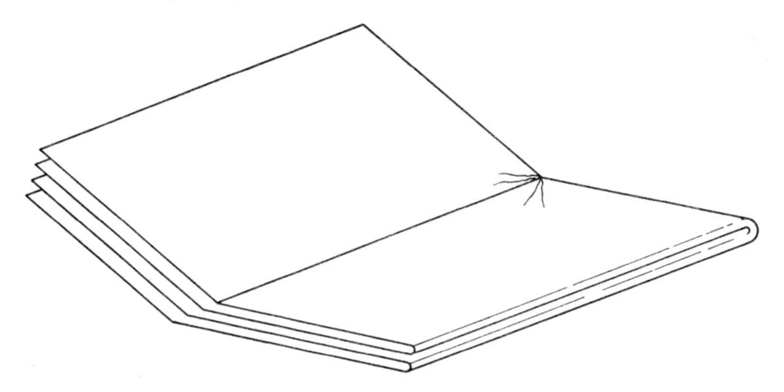

2.1 *Creasing of thick paper at the head of the last fold.*

On average-sized work, approximately 1 000 folds may be completed in any hour. Therefore a 16-page section, *eg* three folds, will be produced at about 300–50 per hour and with a split head bolt (which counts as one fold) about 250 per hour.

When a number of single sheets are collectively folded once, the technique is known as 'lump folding'. This method is used for make-up of account-book sections and no right-angle folds are employed. Very cheap 4-page work, plain covers, etc, may be 'lump folded and pulled'. This implies that a number of 4-page sections are folded together and only lightly creased with the bone folder. These are then laid on to the bench with the fold uppermost and each separate fold 'pulled' with the fingers to complete the fold which is then consolidated with the folder.

Machine folding

On a machine embodying the knife principle the sheet to be folded is laid under timed drop rollers which propel the sheet on to carrying tapes and thence into the folding position. The long edge will be carried up to the front lays and held there by sheet control brushes or rollers. The off-centre sheet will be pulled into position by a mechanical pull-over device and the knife then falls driving the paper between the folding rollers (fig. 2.2). These accept the

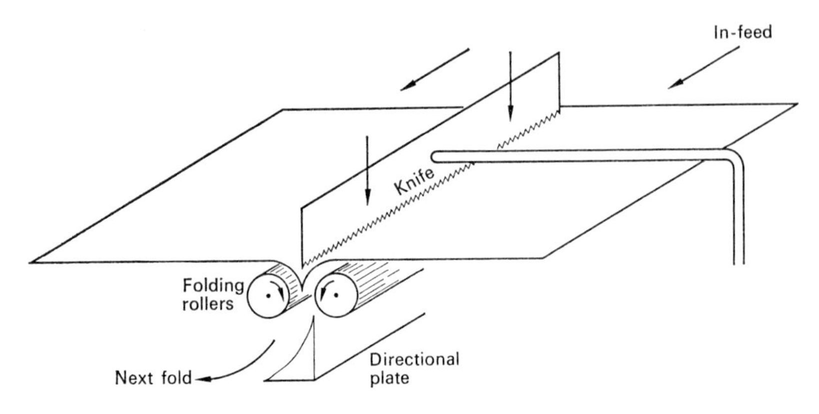

2.2 *Knife folding principle.*

paper, consolidating the fold and passing the sheet to the next fold position. Each succeeding fold is similarly produced and these may be at right angles or parallel to each other. After the first fold position, when there are two or four thicknesses to be folded, it may be desirable to push rather than pull the sheet into its folding position, using either a mechanical movement or a leaf spring.

Whilst the knife principle requires two horizontally placed rollers the 'buckle' or 'plate' method needs three rollers for each folding position. On an 'all plate' machine the sheets do not have to be fed to any particular time cycle and when laid on the feed table are carried into the fold rollers by either tapes or a roller carrier (figs. 2.3 and 2.4). The vertical fold rollers (1 and 2)

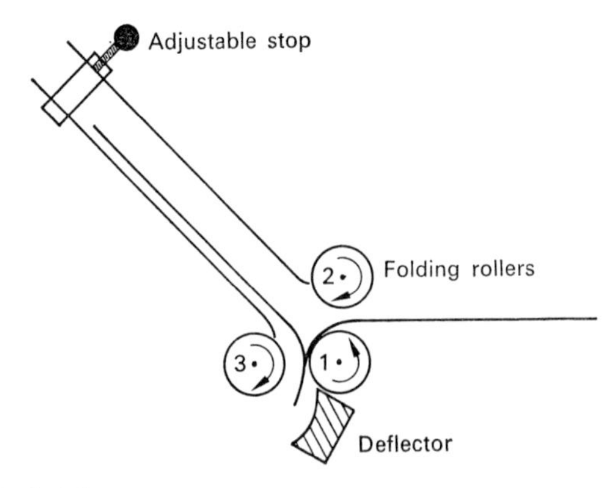

2.3 *Buckle fold principle.*

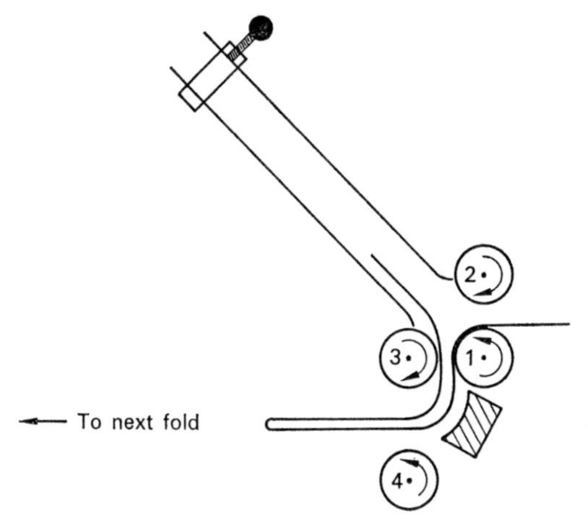

2.4 *Folded sheet progresses to delivery.*

grip the sheet and drive it into the narrow space between the upper and lower halves of the plate until the leading edge strikes the adjustable stop. As the front part of the sheet is now stationary, the folding rollers force the paper to buckle into the space between the leading edge of the plate and the folder

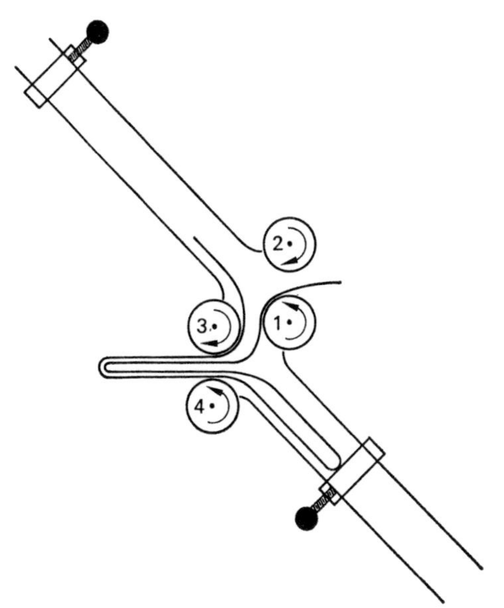

2.5 *Two plate assembly.*

roller. The fold thus made is gripped by the horizontal rollers (1 and 3), nipped, and the folded sheet passed to the next station.

By adding one further folding roller and one plate this assembly can now be utilised to provide two folds parallel to each other (fig. 2.5). If either of the fold positions are not required a deflecting bar can be suitably placed to prevent the paper from entering that particular plate (fig. 2.3).

'Combination' machines have some fold positions using the knife and some the plate principle. All knife machines are mainly used for the production of book and magazine sections. Jobbing machines have several alternative folding layouts embodied, whilst more specialised types may produce sections folded to only one scheme and deliver four or eight sections simultaneously.

All-plate and combination machines are most successful on thinner papers and may be considered more versatile in the type of work which is produced. This may include simple letter and zigzag folds, multi-fold maps and advertising literature with off-centre folding.

Features that may be found on some machines to deal with specific problems are as follows:

Perforating wheels are incorporated on all types to help with the problem of creasing (dealt with under 'hand folding'). These are so placed in the folding sequence that the perforation coincides with the head fold of a section allowing a small amount of movement in the paper and thus preventing creasing in all but the thickest of stocks. Both fine and course perforators are available to deal with differing problems (fig. 2.6).

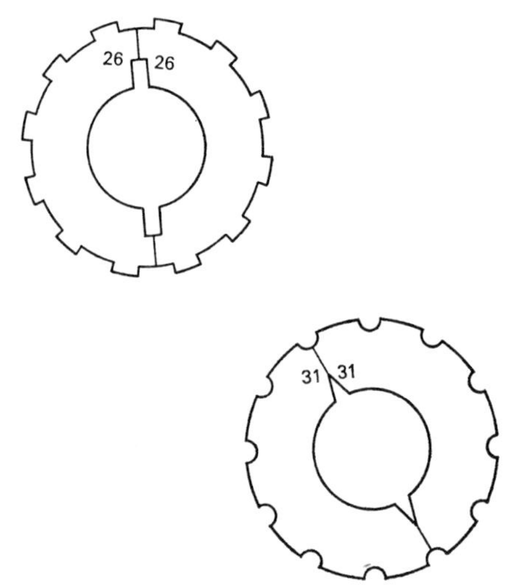

2.6 *Perforator knives.*

Slitter wheels are mounted on the machine to separate work that has been printed in multiple units. This may be high up on the machine between folds or at the conclusion of folding and before delivery.

Creasing wheels may be found on all-plate and combination machines and are discs with rounded edges used to break the fibres of a heavy section or stock immediately before a fold is accomplished.

The ability to control the speed of a machine is often an asset when running thin or difficult stocks and a modern machine will have some form of variable control. This may be either an electrical device to alter the speed of the motor or a mechanical means of varying the size of the pulley on the motor.

One such variable speed system has a motor pulley consisting of a fixed plate and a spring-loaded plate. If the motor is drawn back along its rails so that the distance between the motor and the machine is increased, the V-belt forces the spring-loaded pulley plates apart and then runs on what is effectively a smaller pulley size and the machine speed will decrease. In this way the speed becomes completely variable between the maximum and minimum sizes of the pulley concerned (fig. 2.7).

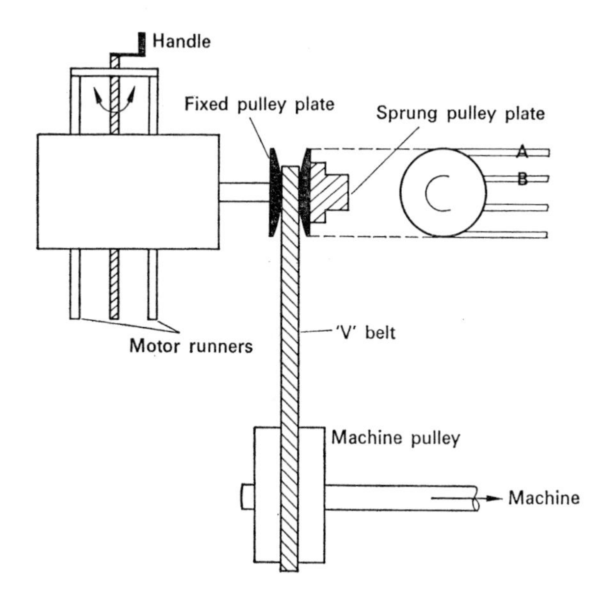

2.7 *Variable speed mechanism. A and B indicate the maximum aod minimum sizes of the pulley.*

Paper feeders

Folding machines may be fed by hand, but increasingly the manufacturers are offering machines that have an automatic paper feeder as an integral part of the machine. This not only relieves the operative of a repetitive task but also makes for higher speeds and more accurate feeding. The wastage rate tends to be higher, however, and where a low spoilage rate is essential, hand-fed machines are used.

The simplest feeding device for small area sheets is the friction or gravity feeder. The work is fanned out and laid on to an inclined surface so that the leading edge of the top sheet is nipped between a friction-driven roller and a driven roller above the sheet. When the trailing edge of the first sheet has passed between the rollers the leading edge of the second sheet immediately presents itself to the rollers, is nipped and fed. Gap between the sheets is the minimum possible and speeds are high (fig. 2.8).

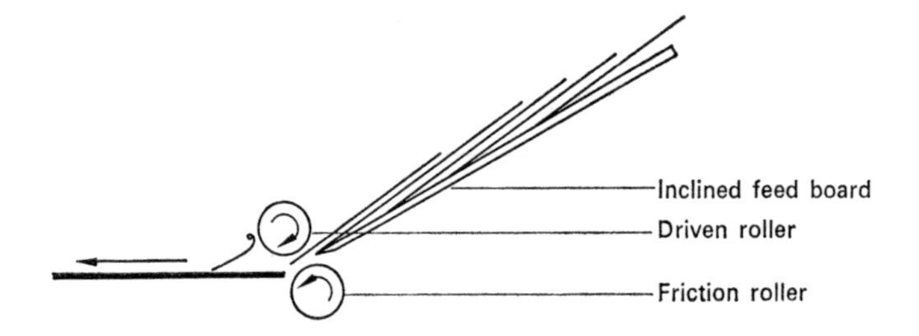

2.8 *Gravity paper feeder.*

Various designs of pile feeder are available. These provide a continuous flow of sheets to the folder with a controlled space between. Actual forwarding of the paper may be by means of sucker cups or by a rotating suction drum; separation of sheets is effected either at the front or back of the pile. This type of feeder is most used on medium-sized machines of both combination and all-plate types. Some feeders require the feed board to be lowered for loading, thus enforcing short breaks in production. Others have the facility to reload the feeder below the existing pile so that changeover time is minimised (fig. 2.9).

To handle larger sheets and to deliver them to the folder at time intervals, a continuous feeder is used. In this context the term 'continuous' implies that the feeder can be loaded at one level whilst the sheets are being fed into the machine at a lower level. In this way the machine can be run for long periods with no down time for feeder loading.

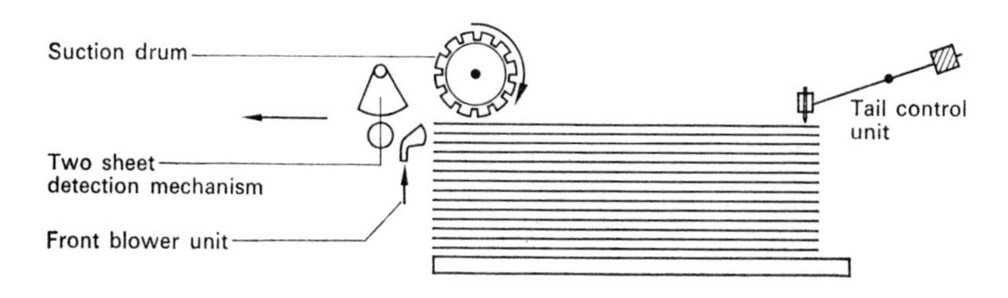

2.9 *Simple pile feeder.*

The sheets are fanned out and banked on the upper board of the feeder and against a gauge. They are carried down the board, around the drum and delivered, right way up for folding, on the lower board and under the 'combing wheels'. These are rotating wheels with small plastic rollers inset into the periphery. When the feeder functions the combing wheels drop on to the top sheet and the plastic rollers skid on the surface of the paper. The friction draws down the top sheet until the leading edge strikes a small spring-loaded finger and through a linkage mechanism the combers are made to rise. At this point the first sheet is about 300mm forward of the bank and the leading edge is under the drop roller of the folding machine and ready to be transferred to the first folding position (fig. 2.10).

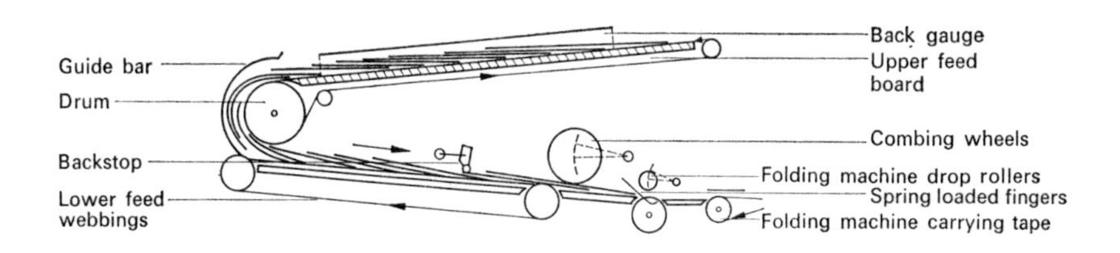

2.10 *Continuous paper feeder.*

Folding schemes

The way in which the folds of a particular machine are arranged will have a direct bearing upon the type of folding or scheme that the machine is capable of carrying out. If, for instance a sheet of paper is passed through two folds arranged at right angles and folding the sheet equally in each direction, then the resulting fold will be described as an 8-page right angle. If the folds are arranged parallel and if the area is equally divided on each fold, then the result will be described as being 8-page parallel. Figure 2.11 illustrates three typical

Description of folds / Typical folding machine layouts	4 page	6 page letter	6 page accordian	8 page right angle	8 page parallel	12 page letter	12 page accordian	12 page right angle	12 page prospectus	16 page right angle	16 page alternative	16 page landscape	16 page two-up	32 page right angle	32 page two-up
All knife (3A, 2, 1, 4, 3, 4A)															
				1				1	1	1			1	1	1
	nil	nil	nil	2	nil	nil	nil	2	2	2	nil	nil	2	2	2
								3	3A	3			3A	3	3
														4	4A
All plate (2, 1, 3, 5 4 6, 7)		1	1	1	1	1	1	1	1	1	1	1	1	1	
		2	2	4	2	2	2	4	4	4	4	2	4	2	
	1	or	or	or other combin-ation	or	4	4	7	5	7	5	4	5	4	nil
	or 2	2	2	of 1, 2, 3	2	or other combination of								5	
	or	3	3	and 4, 5, 6	3	2 from first bank plus one from									
	3					second bank									
Combination (2P, 1P, 5K, 4P 3K) K=Knife P=Plate		1	1		1										
		2	2	1	2	1	1	1	1	1	1	1	1	1	
	1	or	or	3	or	2	2	3	3	3	3	2	3	2	nil
	or	3	3		3	3	3	5	4	5	4	3	4	3	
	2	4	4		4									4	

2.11 *Folding chart.*

folding machine layouts for all knife, all plate and combination machines. A sheet of paper folded by hand in the same manner in which the machine folds will produce model sections having titles listed in the 'description of folds' column. A nil entry indicates that this fold is not normally completed on this machine. Some of the named folds are illustrated in fig. 2.12.

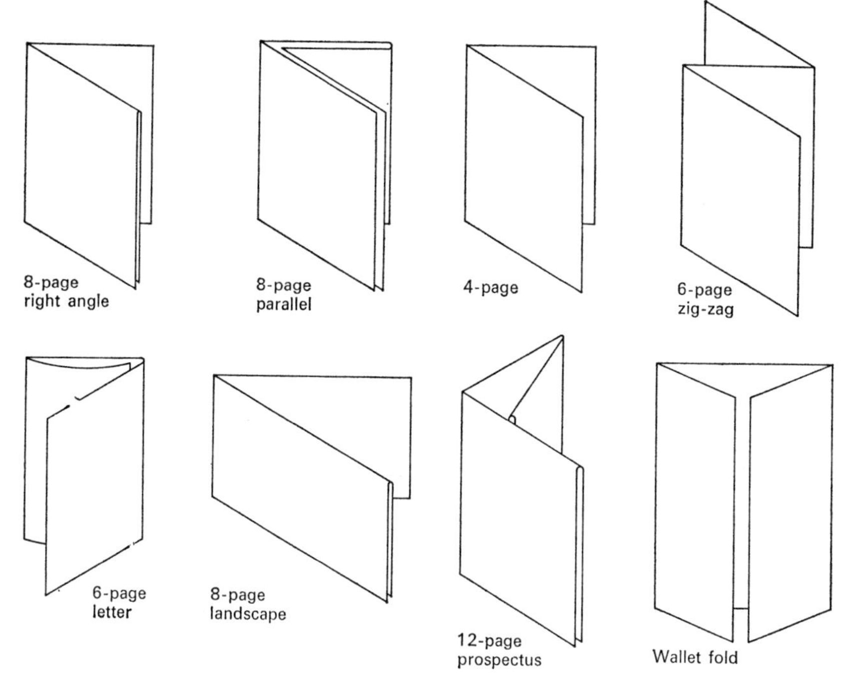

2.12 *Some named folding schemes.*

Folding machine problems

Assuming that the machine is not suffering from extreme wear, particularly in the folding rollers and their bearings, then problem areas can probably be located among the following:

1 Roller pressure setting: excessive pressure will show as indentation of the roller milling on the paper. Insufficient pressure will allow the paper to slip and it may arrive at the next fold position out of time with the folding cycle. Uneven pressure takes the sheet through the rollers at an angle and cuts made by perforators, creasing discs and slitters will be angled across the sheet.
2 Sheet control: this will include the application of dabbers (or slow-down wheels), brushes, balls and wheels at strategic points to ensure that each sheet is being treated exactly alike. Slowing the machine to an optimum speed may also be desirable.
3 Machine direction of stock: this may be parallel or at right angles to the first fold. Buckle folding across the grain of a job printed on stiff paper may introduce errors that are insurmountable.

4 Static: this is a minor problem, but where it occurs the solutions are similar to those for the printing press.

THE MAKE-UP OF BOOKS AND BOOKLETS

After being folded the various units will be combined to make up the job to specification. The number of pages in a section will normally be governed by the caliper stock used. An octavo printed on a bulky stock may look ugly as a 32-page and give rise to protruding pages or 'starts' on the fore-edge if incorporated in a rounded and backed book, but is economic on sewing and folding times. If the same job is imposed to fold as 16-page sections, then the result is aesthetically pleasing but more costly to produce.

Signatures

Two systems for the identification of sections are used. The alphabet in either capitals or lower-case letters is the traditional method with the letters J, V and W being omitted. If the make-up exceeds 23 sections, then the alphabet is repeated in one of the following forms – aa, AA, 2a, 2A – with the remainder of the alphabet following in sequence, and this can of course be further extended as required. The use of arabic numerals is simpler and is steadily gaining favour over the previous method. To indicate an inset the signature cypher is followed by an asterisk while a wrapround will probably show a dagger to indicate its position in the section. Other cyphers that appear at the bottom of the first page of a section may include an abbreviated title, volume number and catalogue number. The following cyphers indicate the third section of the second volume of *The Great Universal Encyclopedia* with the C† wrapping round the outside and the C* being placed inside the centre of the basic section C: GUE II C†, GUE II C, GUE II C* (fig. 2.13).

It is really only essential for the signature to appear on the face of the page and the remainder can easily be transferred to the spine of the section and be hidden by the subsequent binding operations. In some magazine work these cyphers are printed so close to the tail of the sheet that they are trimmed off and in this way the quality of the typography is not impaired.

It will be seen from the foregoing remarks that signatures are necessary in gathered bookwork, especially where the make-up is of a complicated nature, where there are many volumes and where many books of similar appearance in section form are passing through the workshop at the same time. Although not essential in all inset work, signatures are of value in some situations.

Attaching illustrations

Many books and magazines are printed on a paper that allows halftone illus-

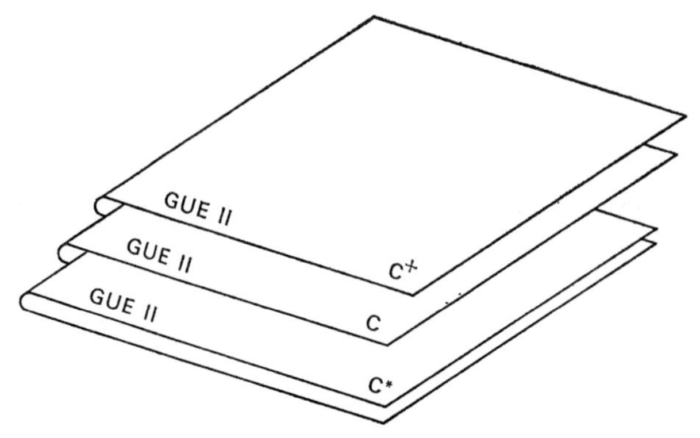

2.13 *Section build-up.*

trations to be included in the text. When there are few illustrations or the book is printed on a rough-textured paper the 'plates' will be added in 2-, 4- or 8-page form.

The following chart shows the principle methods of attachment in commercial work; it must be observed that most of the illustrations will be printed on art paper (fig. 2.14). This material is a base paper sandwiched between

NAME	DIAGRAM	MERIT	DISADVANTAGE
2 page tip-on 2 page tip-in		May be attached by machine Both may be positioned anywhere in a section	Because plates are usually printed on coated papers the adhesion may be weak
2 page tip-on and guard		Very strong attachment	Hand work, therefore expensive
2 page hooked 2 page hooked and pasted		Very strong attachment	Gives unwanted extra thickness at spine Unless pasted, very difficult to handle on sewing machines etc.
4 page inset 4 page wrapround (or outset)		Cheap to complete Strong attachment because it is part of section	Difficult to locate close to relevant text

– – – – Section ——— Plate ||||| Adhesive

2.14 *Illustration attachment chart.*

layers of a chalky substance (the coating), and if an adhesive is applied to the surface with the object of attaching it to a section then the adhesion between the two surfaces may be of only a temporary character. For this reason the attaching of single leaf plates by 'tipping' is far less strong than other methods.

Dividing leaves and boards, single text leaves, maps and charts are other units that may be added to the sections at this stage of the make-up. Methods described for attaching plates are selected to suit the particular need or function.

Maps and charts

Care must be taken when the inclusion of oversize sheets of paper to the book is contemplated. These are usually folded, to a size smaller than the trimmed size of the volume, in a zigzag fashion, and attached by one of the methods previously described. Too many attachments of this type will distort the book and make subsequent trimming and pressing operations difficult.

Alternative methods of including maps in a completed volume are to place them into a pocket glued to the inside of the cover; or to include a section of perforated plain paper into the body of the book which is later reduced to guard size and used to attach the maps by tipping.

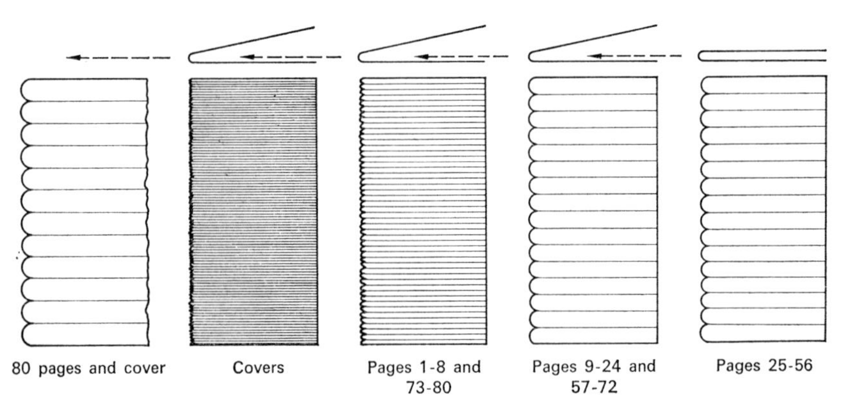

2.15 *Arrangement for insetting.*

Insetting

This is the operation of placing one unit of the book or magazine into another. In this way thicker sections are built up for book work and saddleback magazines are made up ready for stitching. After insetting, the work is knocked (or jogged) up to the head and spine. Where possible the heaviest unit should be placed into the centre of the assembly as these fall more readily than the lighter units, speeding up the operation.

When insetting is completed by hand the sections are laid along the bench in correct sequence, the left thumb opens the work to the centre whilst the right hand carries the section to be insetted (fig. 2.15). Work is often insetted with the head of the section close to the worker, as opening the bolt or fold is easier this way.

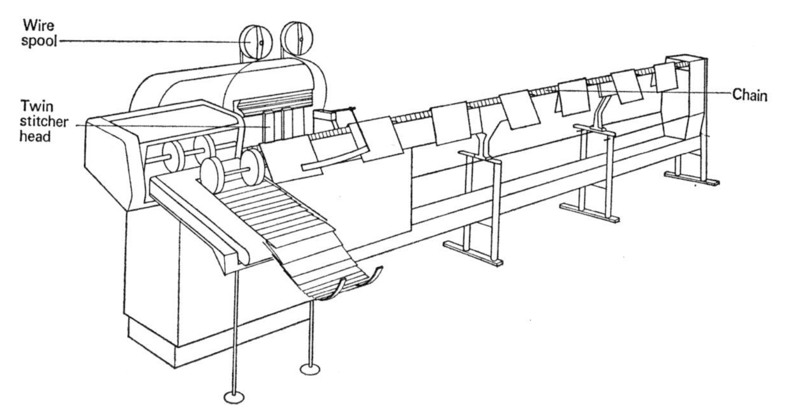

2.16 *Insetter-stitcher combination.*

Machine insetting is normally associated with wirestitching and the resultant machine combinations are referred to as 'insetter-stitchers' or more simply as chain stitchers, with the insetting function implied (fig. 2.16). A continuous chain has a number of feed stops attached to it; the operatives feed a section into each station and as the stops pass, the section is collected and propelled forward, a new section being added at each station. The innermost section is the first to be laid on, the remainder of the sections following with the cover last of all. Transference to the stitching unit is automatic and the completed saddle-stitched booklets are then neatly stacked in the delivery. The latest machines may include self-feeding heads, automatic calipering and rejection devices and automatic transference to single-copy trimmers.

Inserting

In the United Kingdom this term implies the addition of a loose printed slip or document added to a book or magazine after it is bound. This includes the wide range of give-aways, invitations to join societies, book clubs and insurance companies, maps and charts and other ephemera that cannot be included in the book construction.

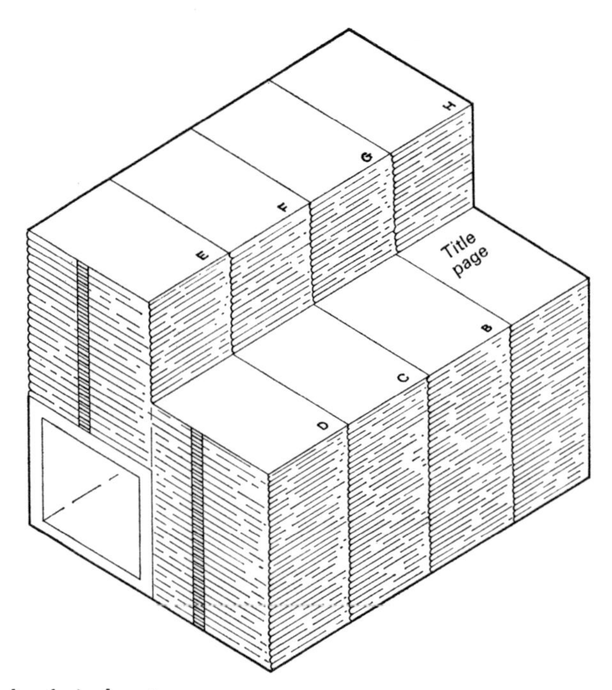

2.17 *Bench gathering layout.*

Gathering

In this process one each of the sections required to make a book are piled on top of each other until the book is complete. In the magazine field an insetted job may be referred to as a 'saddleback' and gathered work as 'squareback'. Much of the gathering in the magazine field is done mechanically because of tight production schedules and longish runs. In the publishers' edition bindery the books are often thicker and have many more sections and this, with the shorter average run, often makes hand gathering more economic.

The sections to be gathered are laid along the bench in sequence and one from each pile is taken until the book is complete. If the job is composed of many sections a double row will be laid out, the back row being raised up, and this will contain the first half of the volume (fig. 2.17). This will also be more productive as the operative is gathering in both directions and is not making an unproductive walk as in single-line gathers.

Many methods are used to speed up this rather slow and tedious process. A simple conveyor belt may have lays attached to it and the sections laid on as these pass the operative. The main disadvantage of this type of device is the extreme length of the conveyors needed to gather thick books. If an operative requires 750mm of working space and both sides of a conveyor can be used, then a unit 7·5m long and more than twenty operatives will be needed to gather a job having 20 sections.

Another simple device sometimes used is a circular table of about 3m in diameter that can be made to rotate. The sections are laid out around the edge in sequence and the operatives are able to sit and gather the sections as they pass.

A mechanism that links the previous methods with the true gathering machine is a series of metal hoppers moving continuously so that the forward and returning halves are in parallel path (fig. 2.18). The speed is variable and the number of hoppers can be ordered to suit the requirements of the work. The machine may be operated by any number of persons and two separate methods of working may be adopted. In the first the sections may be loaded on one side of the machine whilst it is in motion and gathered off the other.

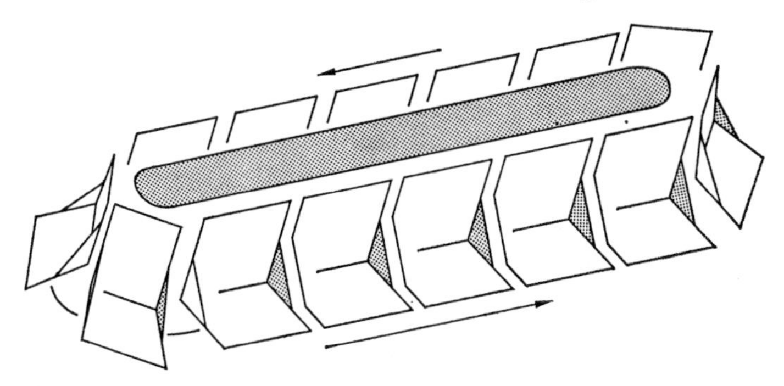

2.18 *Hopper gathering mechanism.*

The alternative is for all the operatives to load whilst the machine is stationary and for everyone to gather when it is running. True gathering machines of the arm and rotary types are dealt with in Chapter 12.

Collating

Traditionally the term collating means the checking of the gathering operation to ensure that the sections are in correct order. In recent years machines have been introduced and termed by the manufacturers 'collating machine' the function of which has been single sheet gathering, and this is dealt with as a topic later in the chapter.

To collate a gathered book the volume is gripped in the top right-hand corner and the left thumb bends the sections downwards allowing one to escape at a time. In this way the signatures can be visually checked section by section. This is a time-consuming operation and only very expensive books are treated this way.

The modern alternative is the use of the 'back step' method of collation (fig. 2.19). A small black mark is printed in the backs so that when folded it appears on the spine of the section. When the book is gathered the back marks appear as a series of steps down the spine and if for any reason a gathering

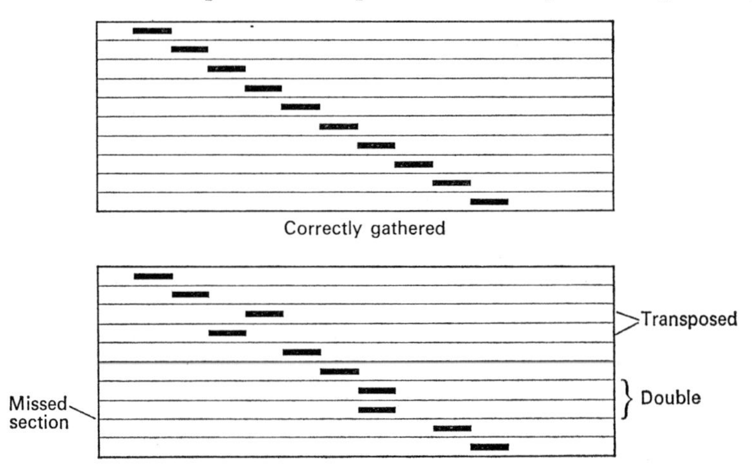

2.19 *The back step method of collation.*

fault occurs, *eg* misses, doubles or incorrect position, it is immediately apparent upon inspection of the spine. That at least is the theory; in practice it is not unknown for the back mark to be misplaced in printing either vertically or horizontally, to be left off altogether, to be folded off the spine or to be obscured by a wrapround plate. If any of these occur the visual effect is spoiled and the system becomes useless.

Because gathering machines are equipped with fault-signalling systems, collating of every book or magazine can be by-passed with reasonable safety. However, a small percentage is checked during the run to ensure that the machine is functioning accurately.

Single sheet gathering

This title is self-explanatory and the process is necessary when the job must be produced in single sheet form, *eg* when the product of the small offset machine has to be bound and, particularly, when manifold work has to be made into sets prior to binding.

By hand the sheets are laid along the bench or (if the work is below octavo or A5 size) in a semicircle around the operative and one sheet from each pile is taken. Some of the paper used in this work is of the thin bank variety,

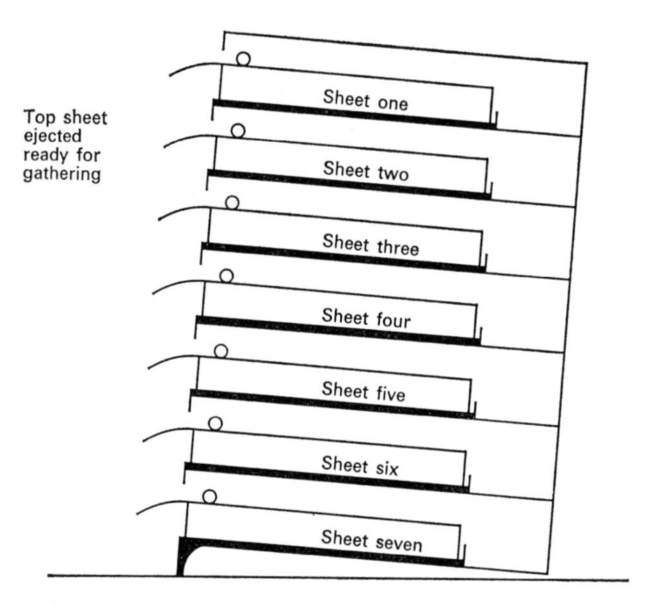

2.20 *Vertical gathering box.*

carbon and other difficult stocks. To assist in knocking up (aligning the sheets, the job may be placed on to the jogger after gathering. This is an angled platform that is made to vibrate by a small electric motor and mechanism. As the sheets are laid they automatically vibrate down to the lays at the lowest corner of the platform.

A wide range of devices are available to assist the operation of gathering single sheets. The simplest is a box arranged with vertical divisions (fig. 2.20). The sheets are placed on to the shelves and extracted one at a time from the top downwards, the operative using a rubber fingerette to assist the separation. Variations include types that push the top sheet forward by either pedal-operated or electrical means. More advanced types are able to handle thin banks, carbons and card substances at high operating speeds.

Manifold work

Receipt, delivery, order and many other commercial functions require printed forms in multiple sets. These may be printed or ruled in single leaves and treated as above, or if a sewn and flat opening book is needed, then the book must be built up into section form, and is limited to duplicate work.

In this case the paper is cut to twice job size and the first sheet (printed pages 1 and 3) perforated (say) 15mm either side of the centre fold position.

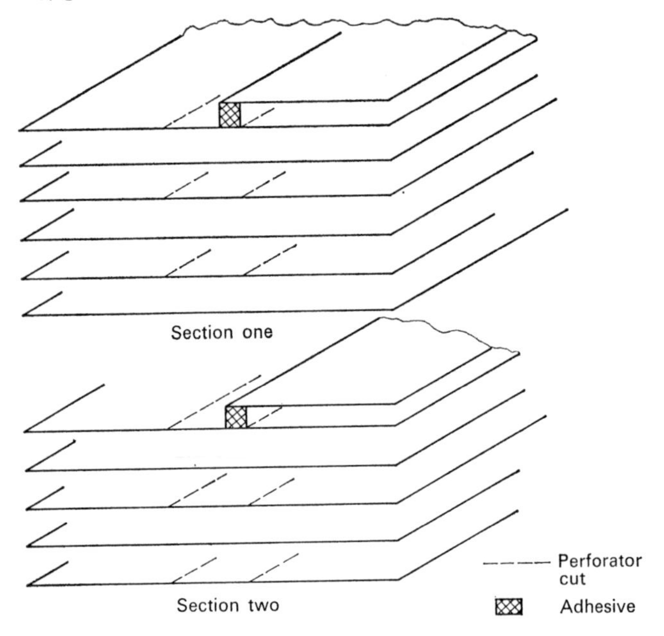

2.21 *Sequence of sheet interleaving for sewn duplicate work.*

The second sheet (often a plain or coloured sheet) is left non-perforated. These are then gathered in alternate sequence and folded in sections of suitable thickness. The first section will have the plain sheet outside the section while the second section should be folded with the perforated sheet on the outside and this sequence will be repeated throughout the book. To restore the alternating sequence of the pages in the centre of the sections a single leaf is tipped in with paste. The book is now ready to be bound in a suitable style (fig. 2.21).

Other constructions, *eg* triplicate, quadruplicate, quintuplicate and sextuplicate are made up by single-sheet gathering and must be side-stitched or adhesive bound.

Modern production methods utilise web-fed rotary printing machines that issue the work one, two or three up, printed, numbered, perforated, punched and gathered ready for the binding process.

3. Miscellaneous operations

This chapter deals with a number of more or less unrelated processes that may be used in the production and finishing of paper products before or during the make-up stage. Reshaping paper by cutting, drilling and punching is fundamental to the finishing process and various applications are to be found in many of the different specialisations.

Punching

This usually implies the making of one or more holes in the job in a predetermined position. Although hand-held and table-mounted devices are

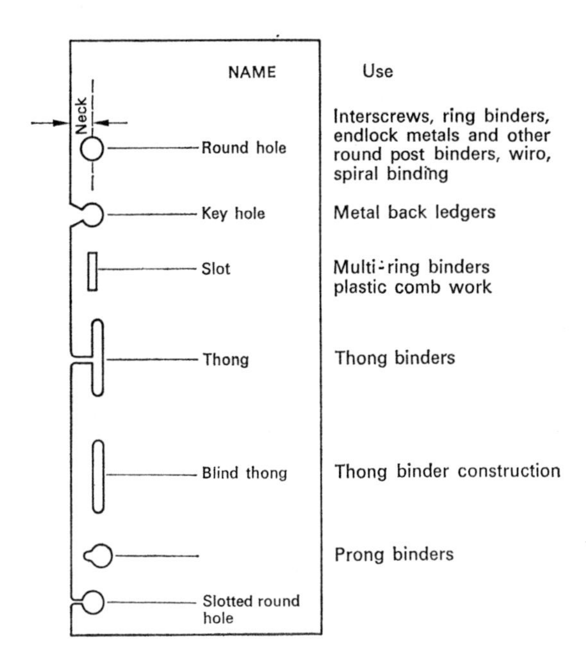

3.1 *Punching shapes and names.*

available the trade normally utilises floor standing and foot operated or foot controlled power operated machines. On this type of machine the sheet is laid up to front and side lays and the treadle pressed. This forces the punch bar to move downwards carrying the male punches through the paper and into the female punch plate underneath. The number of sheets that may be punched

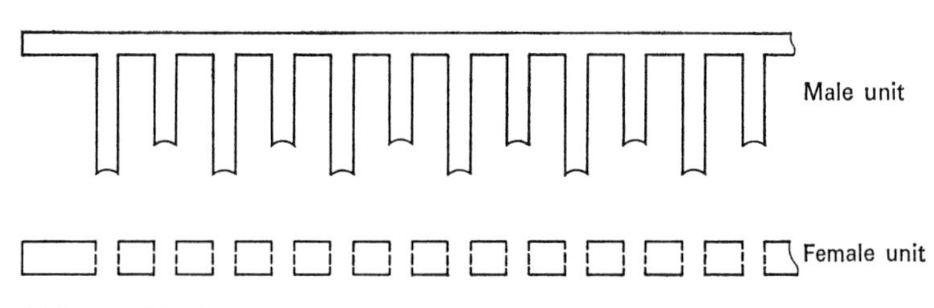

3.2 *Multi-punch head.*

simultaneously is governed by the quality of the stock being punched and the power of the machine used. Empirical methods are used to arrive at the number for a particular job. The main shapes and their uses are illustrated (fig. 3.1) but any shape can be punched providing a tool can be made.

The simplest machines have individual punches that can be positioned to suit varying requirements. In this way the number and type of punches carried on the punch bar at any one time can be varied to suit the job. When there are considerable quantities of sheets to be punched with many holes it is often worth while to equip the machine with a multi-punch unit (fig. 3.2). An example of this is punches for multi-ring binders requiring either round holes or slots 8mm apart.

Perforating

This is in effect the punching of a series of small holes very close to each other so that a portion of the sheet of paper may be readily torn away. Three methods are described as round hole, slot and slit (fig. 3.3). A bar perforator has a number of small round punches (pins) mounted in groups or units and these work vertically through a female punch bar. This type of machine is used mainly for miscellaneous work and short runs. Stop perforating is achieved by removing a pin unit and aligning the end of the shortened pin bar with the stopping point on the job.

Both slot and round hole perforating may be produced on rotary machines. The round pins are mounted on a rotating wheel and work into a female

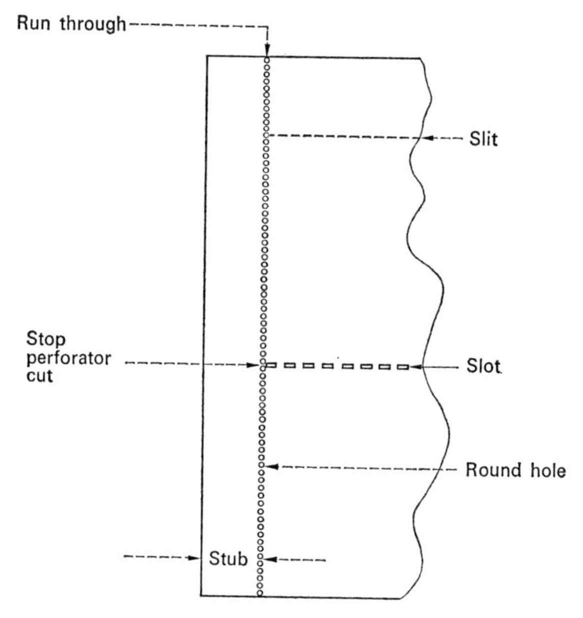

3.3 *Perforator styles.*

punch wheel rotating underneath. A sheet of paper running between the two rotating wheels is perforated in a straight line. As the wheels are of large diameter, stop work may again be produced by withdrawing pins. The smaller discs of the slot type, work into a slotted wheel beneath and have to rise and fall to accomplish stop patterns. The lifting movement may be controlled by a lifting cam action or by a photo-electric cell. Slit perforating rule is incorporated into the printing forme and is not normally a separate binding operation.

Drilling

Round holes are more quickly produced by drilling than by punching and machines are available with one or more heads. The drills are of hollow construction and the waste moves up through the drill and is delivered into a bin. Diameters of 2·5mm up to 12·5mm are usual and the maximum amount of paper drilled at one time may be from 25mm for small diameters, to 60mm for larger diameters.

Some machines have a moveable side gauge and the paper is moved on to them for each drilling position; others have a table shift mechanism where the whole table moves on to the new drilling point. Power to drive the rotating drill through the paper may be derived from a hand lever, a foot-operated treadle or a simple hydraulic ram. Machines may have the facility to control the speed of the electric motor that gives the rotational power to the drill. This is sometimes helpful if large diameter drills heat up and anneal the working surface when run at high speeds.

Gauge bars of given dimensions are sometimes used for repetitive work; these are particularly useful when, say, twenty holes are being drilled in paper going into a multi-ring binder or visible index ledger. Greater accuracy can be obtained as well as a speedier set-up and the ability to return to exactly the same setting is an advantage.

Providing the rotational motor is not switched on, these machines can often be used as punching machines and other facilities available on the machine may include round-corner cutting, thumb-hole index cutting and punches for converting round holes into slotted and keyhole shapes.

Index cutting

Indices that are cut into the body of the book (address books and bound ledgers) are described as 'cut indexes' and the cutting may be a hand or machine operation.

By hand the length of the book is divided into the number of letters to be

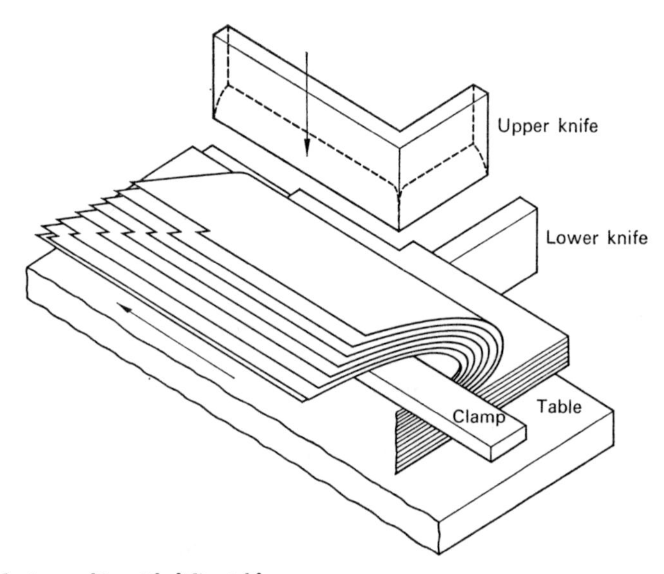

3.4 *Indexing machine with sliding table.*

used, normally 20, 22, 24 or 26. The first sheet is cut away so that a square of paper is left attached at the top of the page; the second sheet has an area equivalent to two divisions left and so on until the index is complete. If the whole book is being cut through, the number of leaves in the volume is apportioned to each letter of the alphabet according to its use. Suitable tables appear in BS 1544.

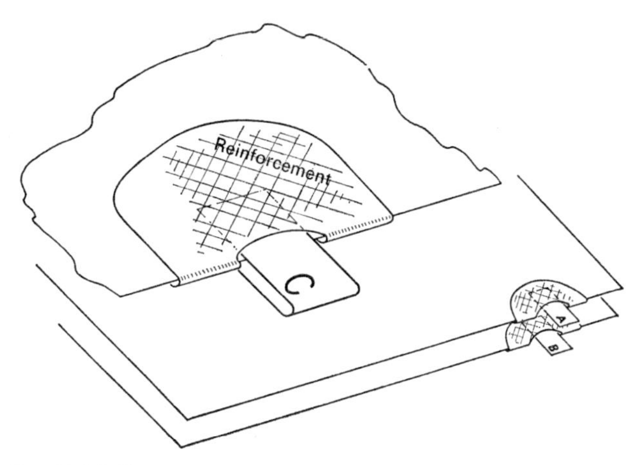

3.5 *Reinforced tab indices.*

A simple machine is one that has an L-shaped knife that can be set to cut in even divisions. The book is assembled, face downwards, under the knife which is on the left-hand side of the slide. The 'Z' leaf is removed from under the knife and a cut is made, the knife immediately sliding right to its new position. The 'Y' leaf is removed and another cut made and so on to completion (fig. 3.4). The vertical action of the knife may be controlled by a foot-operated treadle or may be power driven. Other machines have table shift mechanisms and may print in the letters on the tabs. A cut-through index appears very clumsy on a thick book and a neater appearance is obtained on letterpress books of reference by having a semicircular thumb cut. These may be cut to coincide with printing on the fore-edge of the leaves or shaped pieces may be pasted in afterwards.

Tabbing of index cards is completed by punching the individual tabs in groups and then gathering into sets as a secondary operation. Reinforcing of cut and tabbed indices is done by strengthening the area with paper, textiles or film (fig. 3.5).

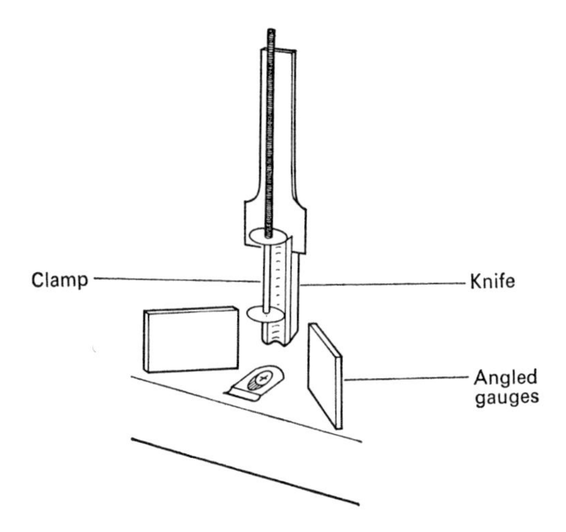

3.6 *Round cornering machine.*

Round cornering

To prevent the wearing and dog-earing of square corners of books and cards the corners can be given a slight radius (fig. 3.6). The shaped cutter is mounted in a hand-, treadle- or power-operated machine. The work is placed in angled gauges and the cut made. Some drills, punches and indexing machines have attachments to do this operation.

Label punching

Labels having 90° corners may be cut on a power guillotine but those of irregular shape will need to be punched out with a forged steel cutter.

Two distinct methods of label punching have been developed. The older utilises a heavy power press with a reciprocating head and push-in pull-out type of bed. The paper pile, about one inch thick, is arranged on the bed and the cutter placed over the label to be cut. The bed is then pushed in and the reciprocating head comes down and forces the cutter through the paper (fig. 3.7). The bed is withdrawn, the cutter lifted out, emptied and the process repeated. Problems of split-out occur when the thick cutter is forced through the paper and this limits the number of sheets that can be punched simultaneously.

The second method requires that the labels be accurately guillotined so that the label is centred on the resulting rectangle. These are then placed on the ram of the power machine that forces the paper stack on to the fixed

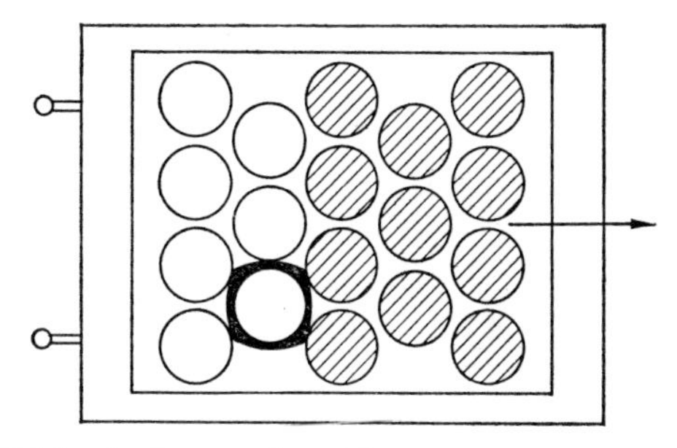

3.7.1 *Plan view of platen showing cutter in position.*

cutter (fig. 3.8). The cut labels pass right through the cutter and appear on the delivery slide on top of the machine. Although this method requires that an extra cutting operation is performed the output of the machine is very high.

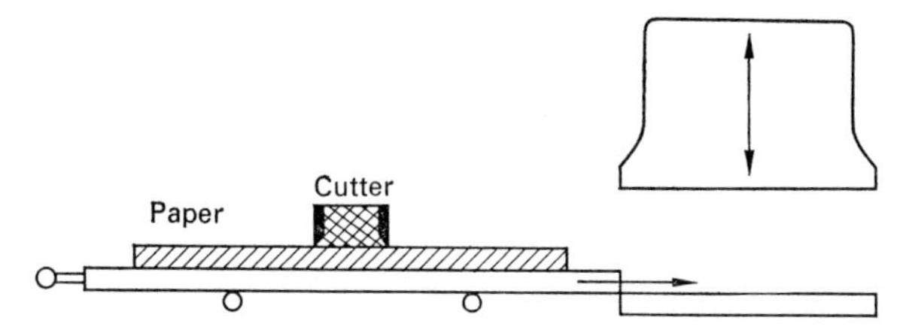

3.7.2 *Cross-section of label punching machine with reciprocating head.*

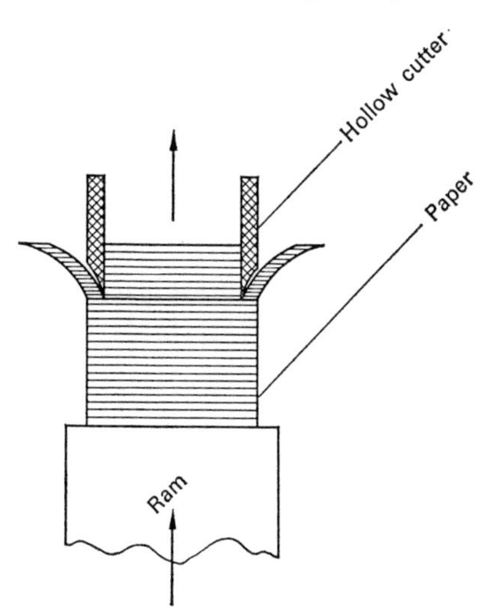

3.8 *Ram type of label punch.*

Creasing

In this context the term creasing implies the breaking of the fibres in paper or board to facilitate folding or bending. This may be necessary during the make-up of showcards, calendars, covers and indeed the finishing of any thick or stiff materials that must be folded or bent neatly.

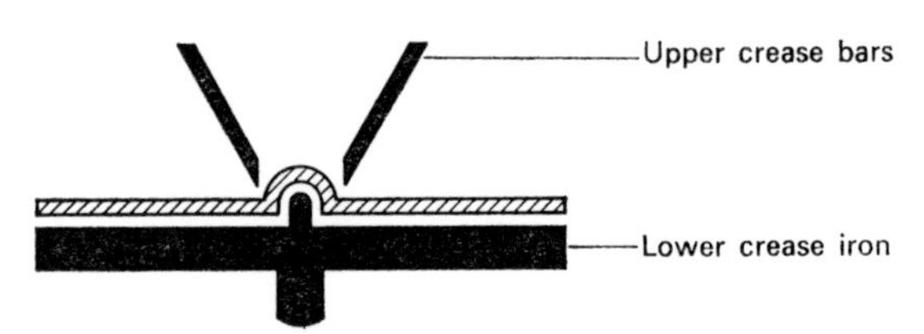

3.9 *Creasing principle.*

Simple mechanisms are based on male and female crease bars of different thicknesses, depth and spacing. These machines may be hand or treadle operated and may be adjusted to give different effects (fig. 3.9). When folded over, the creased 'bump' should appear inside the fold.

In the case of certain loose leaf sheets it is necessary that the binding edge of the sheet should be flexible and if required the paper is passed through a roller mechanism having a series of creasing positions (fig. 3.10). The term

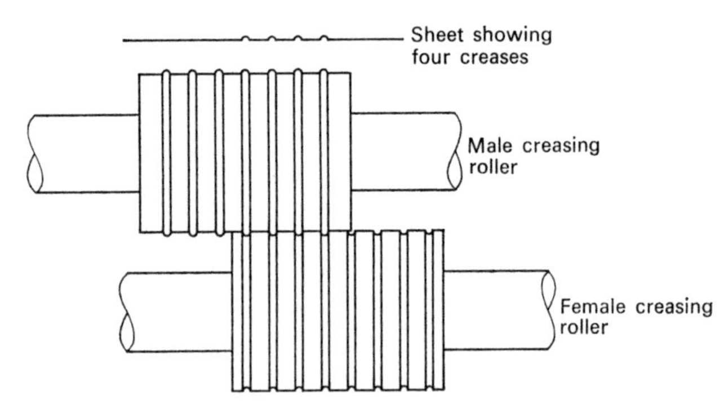

3.10 *Creasing device for loose-leaf sheets.*

'score' usually implies that the bending quality of a thick card has been improved by partially cutting through the fibres at the point of the bend.

Numbering and paging

Many commercial books and forms require to be numbered in a particular sequence for later ease of reference. Simpler work and longer runs are numbered by a numbering box being included in the printing forme and in the case of rotary printing machines special numbering heads are incorporated on the machine.

Small quantity and miscellaneous work is usually numbered in the finishing department using a hand-held machine. These are similar to office machines and may have six numbering wheels giving the largest number as 999999. Most machines have the facility of numbering in the following sequence:

1 Repeat, *ie* the same number at each operation.
2 Continuous, *ie* the addition of one number each time.
3 Duplicate, *ie* two numbers identical before moving on to the next number.
4 Triplicate, *ie* three numbers.

and so on up to six numbers.

Some machines have space to add an inserted prefix letter if required. Most commercial work is numbered on the faceside only and girl operatives achieve high outputs with these hand-held devices, maintaining surprising accuracy by visual means.

Rotary numbering machines are available that have one or more open cylinders on to which numbering boxes can be fitted and these are widely used for numbering sheets of printed forms.

Account books and ledgers are numbered during or after binding according to house practice. The numbers appear in the top left- or right-hand corners and if page 1 commences on a left-hand page and page 2 on the right-hand page the book is said to be page numbered (or more simply 'paged'). Certain books of account call for both sides of the opening being numbered identically and this is said to be 'folio'd'. The nomenclature of the books, the rulings used and how these should be numbered are discussed in BS 1544.

To speed the production of this class of work, one or two numbering heads are mounted on a swivel, the book laid on the table beneath and the numbering impressed by pulling down a handle which levers over the head. This type of machine can produce all the numbering sequences previously mentioned and can also 'skip' number, *ie* jump a number as required. This is necessary when a book is being 'paged' and the right-hand (recto) leaves are processed first and the machine is then reset for the left-hand (verso) half of the book. Treadle versions of this machine are also available.

Gumming

This is another simple process that may have to be completed during the make-up of the job. Some forms, *eg* receipts, are supplied to the customer in

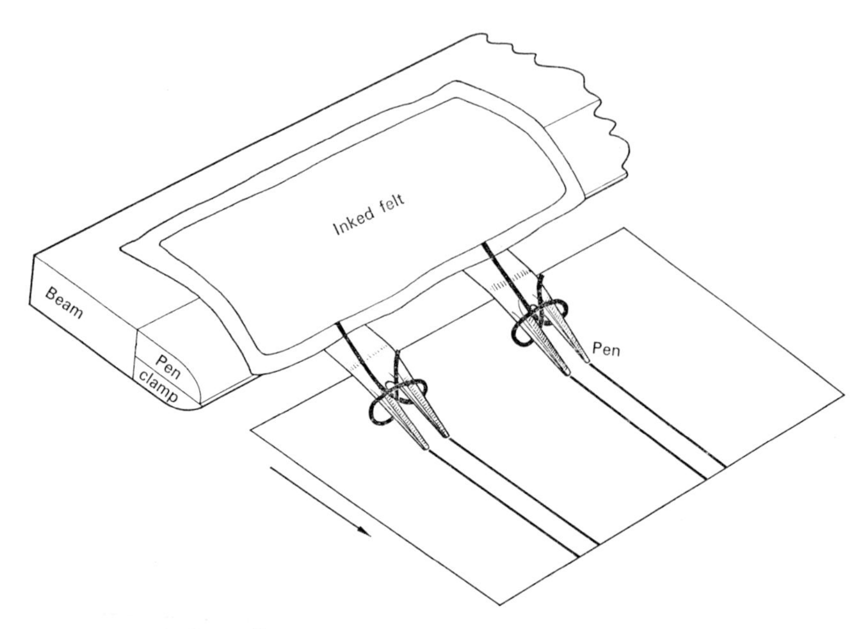

3.11 *Principle of pen ruling.*

books, one, two or more to view and with one of the underside edges gummed for attaching to other documents. Car stickers and window bills are other examples of strip gumming requirements.

The gums used are usually gum arabic or prepared formulations direct from the adhesive manufacturers and these are considered more fully in Chapter 14. By hand, the leaves are fanned out or masked and the gum brushed on. The sheets are then laid out on drying racks until dry enough to be picked up.

Larger sizes and quantities may be strip gummed on a small sheet gluing machine adapted with gumming wheels for the purpose (see Chapter 10). Drying may be speeded up by passing the sheets under a heated hood on a conveyor belt.

Machine ruling

Many commercial books of account, students' notebooks, documents and forms require guide lines to assist the user to space and align the items entered. Often no other imprint is required on the sheet and the simplest method of putting these lines on the paper is by the use of the pen or disc ruling machine.

The pen method of ruling utilises pens of specially shaped brass mounted in the required sequence on a beam set across the width of the machine (fig. 3.11). Felt pads laid on top of the beam are charged with liquid ink and this is led to the individual pens by wool strands and thence down the groove of the pen to the working face or foot. The paper is laid on to a carrying felt which leads the paper beneath the pens allowing them to draw a firm line of the width of the pen and in the colour of the ink being carried.

The horizontal lines in ledgers in a particular shade of blue are described as 'feints' and these are run through, *eg* they continue from edge to edge of the paper. A number of beams may be used on one machine and, by a system of adjustable cams operating on the end of the beam, may be made to rise and fall independently, producing a line called a 'stop pattern' (fig. 3.16).

Timed feeding by hand is achieved by a simple gate mechanism or the machine may be mechanically fed. Tensioned threads are used to control the moving paper on the felts and these are adjusted to run between the points of ruling. After ruling, the felts and threads carry the sheet in a flat 'Z' to the delivery. In this way the ruled paper is made to traverse almost three times the length of the machine before delivery, allowing natural drying of the ink to occur. A blotting roller removes accidental excess of ink before the sheet is discharged into the lay-boy or delivery box (fig. 3.12).

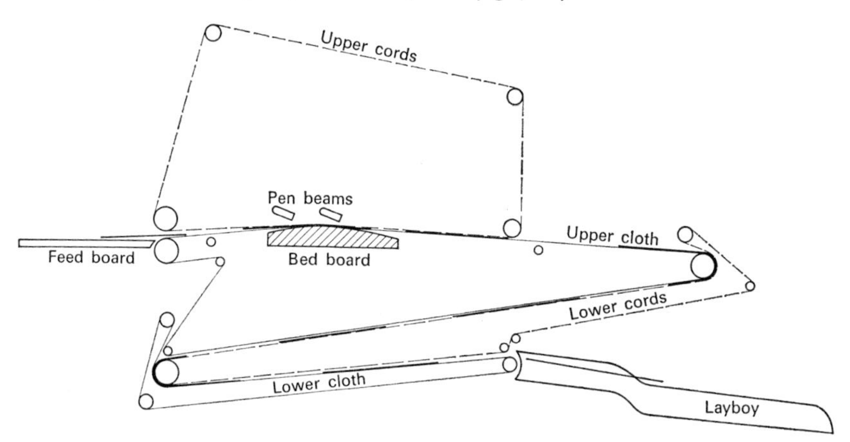

3.12 *Pen ruling machine showing progression of paper from feed to layboy.*

The pen machine tends to be a process requiring good craft experience and a delicacy of touch to overcome problems of complicated patterns, curled and heavily sized papers. Low capital cost and short set-up time may make the process economic for short run work but the productive speeds tend to be low.

In the alternative disc ruling technique a number of brass discs are mounted on a shaft, charged with ink from a rubber roller and allowed to run on the paper. The width of the face of the discs and the spacing collars between them are measured with the printers' point system. The design and progression of the paper through the machine is similar to the pen except that whereas the

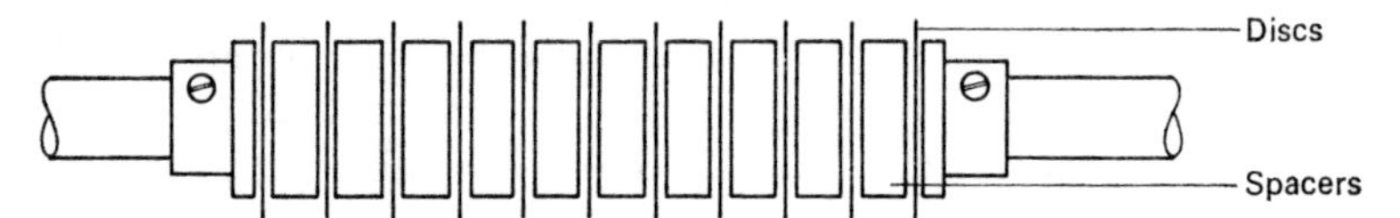

3.13 *Shaft showing discs and spaces.*

pen works against a flat surface the disc shafts are arranged around an impression cylinder (fig. 3.14). With two cylinders mounted in tandem and a suitable arrangement of felts and cords both sides of the paper may be ruled with one run through the machine.

These machines are usually automatically fed; the set-up time is rather longer than for the pen version but the high speeds of production obtained make for a cheaper product. It is therefore ideal for producing the longer runs of straightforward rulings on commercial papers.

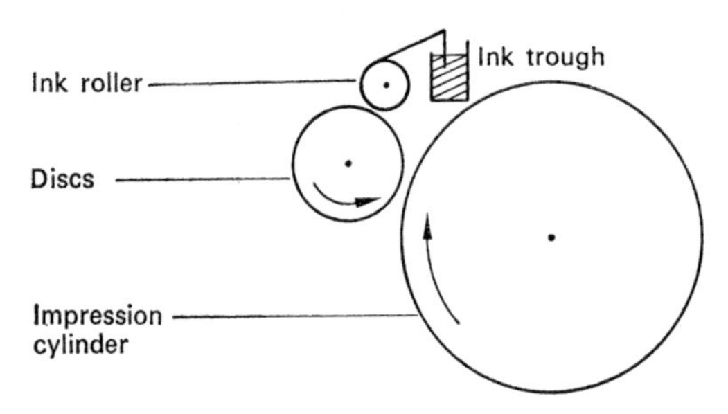

3.14 *Inking system on disc ruling machine.*

Both pen and disc machines may be adapted to rule in two directions, *eg* feints and down lines with one run through the machine. This is in effect two machines in 'L' formation with a sheet transfer device at the junction. The first leg rules the feints in one or more colours and delivers the sheet to the transfer mechanism which redirects the sheet on to the second leg where the down lines are completed. These machines are highly productive and their use is only possible when very large quantities of identical rulings are required.

Web-fed machines are found in some highly specialist applications; a

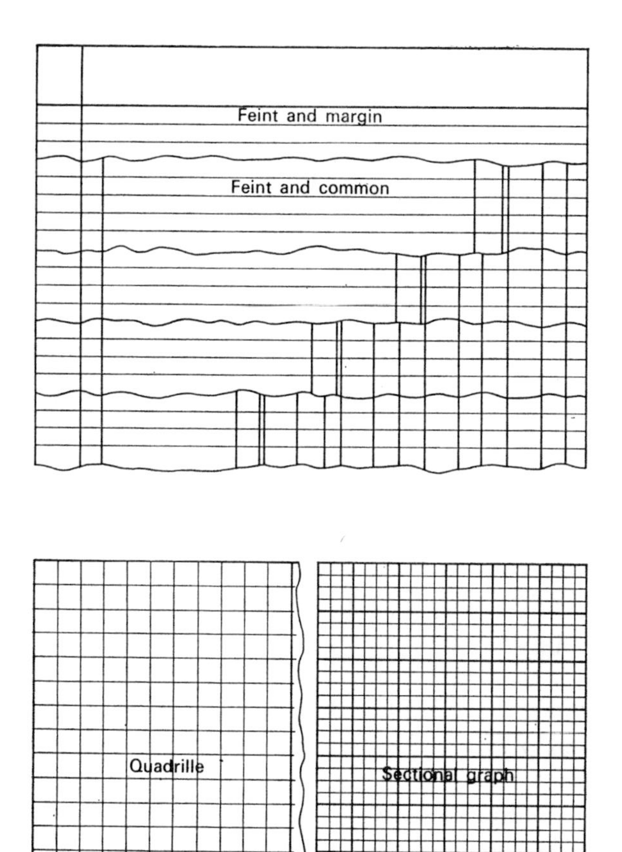

3.15 *Run through ruling patterns.*

typical web-fed disc ruling machine rules feint lines, prints cross lines from rubber stereos, sheets the paper to size, adds preprinted covers, folds once and saddle wirestitches student exercise books at high speed. Disc ruling units are also to be found on web-fed rotary printing machines putting in the coloured lines on the web which is subsequently printed and folded into sections for diaries, address books and similar products.

Standard ruling patterns may be divided into two main groups; common and stop designs. Some standard common patterns are described in BS 1544 and it should be noted that both the horizontal and the vertical lines are run through. These patterns are particularly used on forms and documents, cheap flush books and common account ledgers (fig. 3.15).

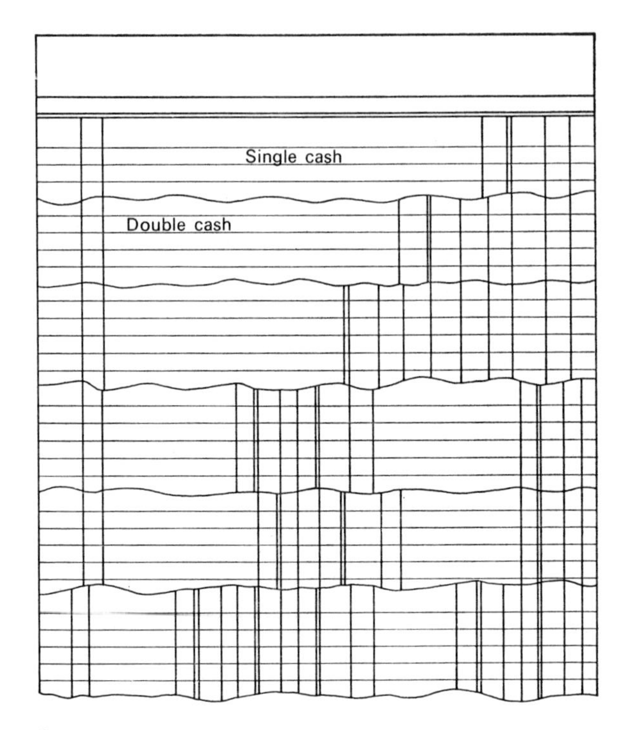

3.16 *Stop ruling patterns.*

On legal documents and better-quality ledgers the horizontal lines terminate at the head line of the page (fig. 3.16) and many commercial users specify non-standard rulings. These may include such features as 'box heads' in which headings are printed, oblique heads, vertical and horizontals that do not reach either edge of the sheet, multi-coloured work, etc. In certain circumstances it may be convenient to rule a job more than one copy up, but with a complicated ruling pattern the set-up time and rate of spoilage may both be high. In recent years the application of offset printing to the production of forms and documents has considerably reduced the number of machine ruling installations.

4. Securing operations

The fastening together of printed leaves or sections prior to binding may be completed in one of two basic ways. The simplest is the process of stitching and may be defined as the completion of the fastening with one revolution of the machine. Wire and thread stitching come into this category.

The alternative method of sewing requires many operations to complete the joining of the units of the volume either by repetitive hand methods or by repeated movements of the machine.

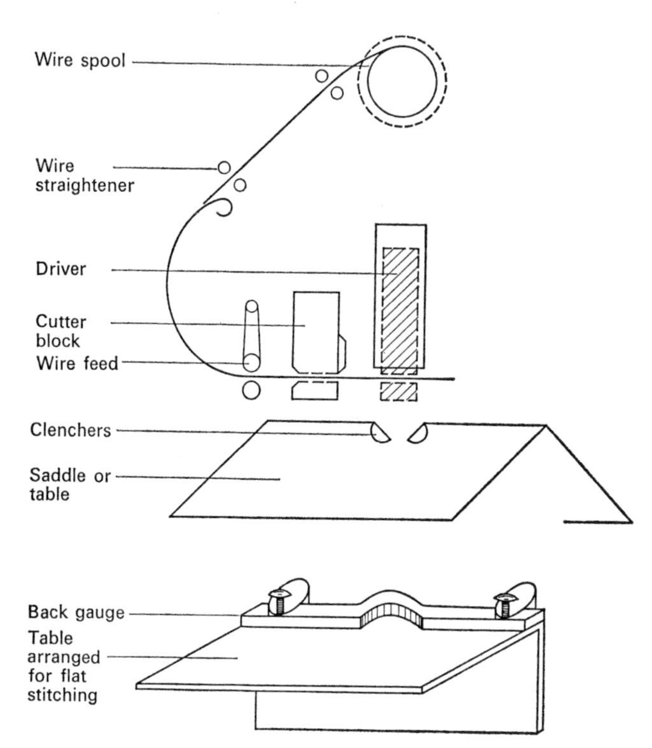

4.1 *Elements of a wire stitching machine.*

Wire stitching

The cheapest and most widely used method, it closely resembles stapling as used in commerce. The stapling machine is loaded with performed staples whilst the wirestitcher commences with a reel of wire and performs the stitching operation at very high speed.

A typical machine (fig. 4.1) accepts wire from the spool and passes it through a straightening device so that the leading end presented to the cutting and shaping mechanism is free of curl and kinks. In some machines this aspect is particularly important and of course is most difficult with thin wires. The wire is fed into the shaping mechanism and held whilst the correct length for the job is cut off. The legs of the staple are then formed by bending the wire down either side of the bender block. The bender is withdrawn from the completed staple leaving the driver free to drive it into and through the paper. Beneath the work being stitched two small moving parts bend the legs of the staple inwards to close or clench the stitch (fig. 4.2).

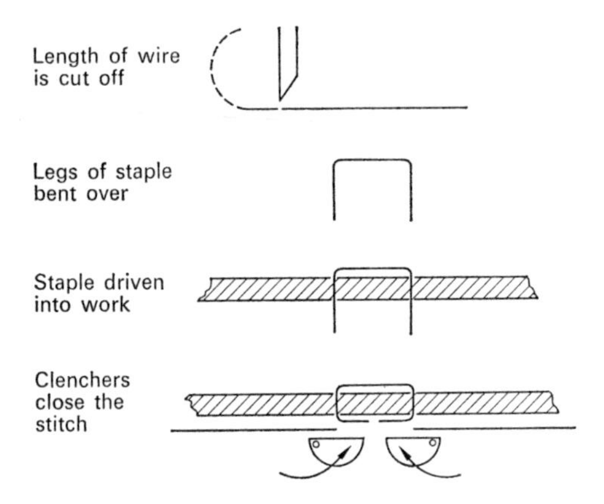

4.2 *Stages in stitch formation and clenching.*

Stitching wire is usually of coated or uncoated steel. The coatings used are copper, plastic or aluminium and may be considered decorative and protective. For every-day work wire having a round cross section is used in nine basic calipers between 0·4mm and 0·9mm. In situations requiring a wire of considerable penetrating power but having a minimum bulk the wire used is described as 'flat'. In fact this is a round wire that has been passed between flattening rollers giving the wire two parallel flats. In this case two measurements are required, *eg* 0·6/0·45mm; the flattened surfaces being the thinner caliper.

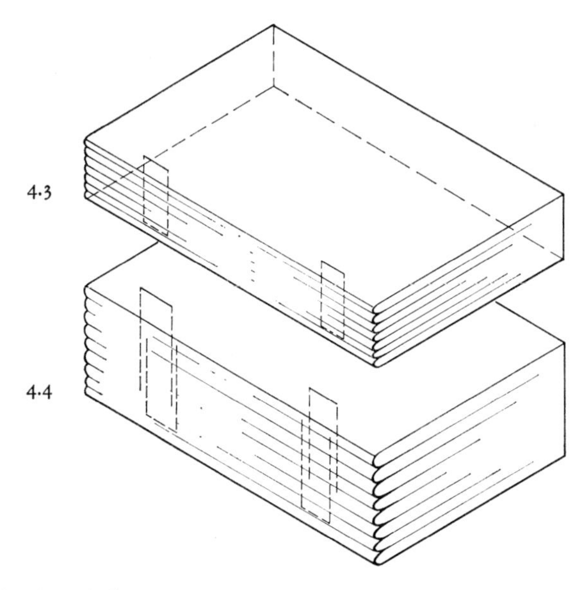

4.3 *Book side stitched.*
4.4 *Book stabbed from both top and underside.*

The work table or saddle of the machine may be adjusted to lie horizontal or in the shape of an inverted V as required by the job in hand (fig. 4.1). Suitable gauges are provided and the machine will have a protective, or better still, an interlocked guard.

A foot treadle is the method of starting the mechanism and the clutch is so arranged that the machine will continue to stitch until the pressure is removed from the treadle. In this way a competent operative can 'run on', *ie* put two or more stitches into a job by moving the work along the table without removing the foot from the treadle. Double headed machines are available but seldom used.

Work may be inset and saddle stitched or gathered and side stitched. A stitch, by definition, is clenched (fig. 4.3) but some machines are adjustable to pass the staple into the paper without clenching and this is usually referred to as 'stabbing' (fig. 4.4).

Stitches which are accurately positioned are said to be 'placed' (fig. 4.5) and there are arguments for and against this practice. Spreading the stitches out over a wide area (fig. 4.6) prevents the wire causing extra bulk (or swell) in the spine, and this in turn makes the subsequent trimming operation easier to control. In extreme cases, *eg* thin 8-page sections, it may be desirable to

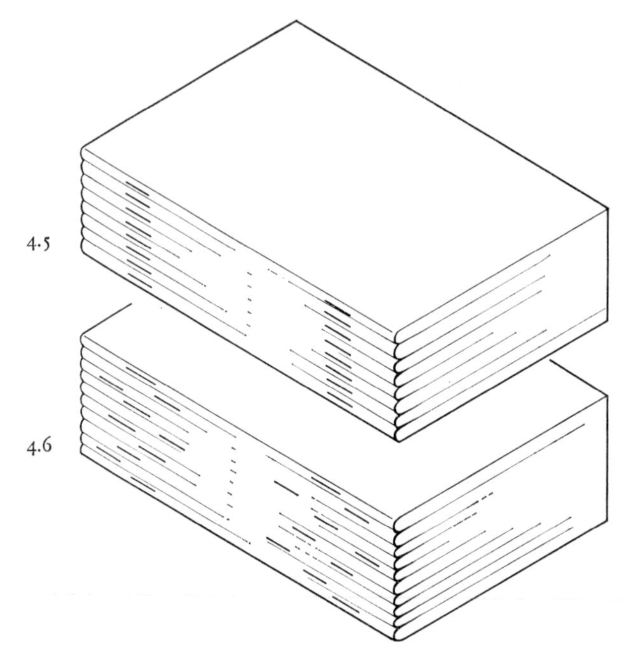

4.5 *Saddle stitches 'placed'.*
4.6 *Saddle stitches 'staggered'.*

trim the fore-edge and tail of the job before the stitching operation is completed and in this way prevent the stitches bursting through the paper during the trimming process.

Thread stitching

In the production of small runs, special orders, educational work where wire is unacceptable or where rebinding quality may be of paramount importance, thread is preferred to wire. By hand a three- or five-hole stitch may be executed with the fastening knot inside or outside the job as required (fig. 4.7). To make the holes in a thick side-stitched job a stabbing device is employed.

A thread-stitching machine is fed with a reel of cotton thread or coloured cord which is fed across the machine and held by retaining clips. The job is placed up to a gauge on the table and the head lowered to hold it firmly in position. Three strong hooked needles move up through the table and the paper, locate the thread and pull it downwards into three separate loops. A shuttle working from the right and through the first two loops, grips the end of the thread and draws it back through the centre loop to the right-hand end of the machine. At this point a knotting mechanism operates, the thread is severed and the new end relaced before the machine comes to a halt with the head open.

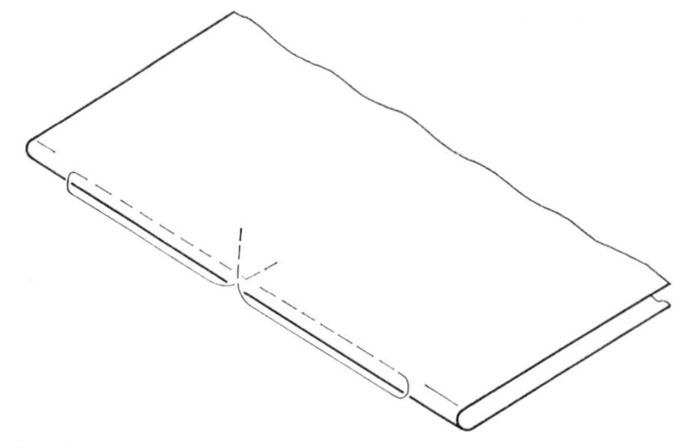

4.7 *Three-hole saddle thread stitched.*

These machines can be adapted for saddle and flat stitching, calendar looping and similar tasks using a variety of threads and coloured cords in two stitch lengths. This type of machine can be difficult to operate smoothly and production speeds are rather slow.

Thread sewing

Hand or machine sewing is employed when the book is subsequently to be bound or cased. The choice of sewing method used will depend upon the style of binding and this is discussed in Chapter 12. Hand sewing is an expensive process and is only employed when a suitable machine is not available, the run is very short or the style and subsequent cost of the completed binding requires a hand-sewn book unit.

'French' sewing is a method that uses only thread and has no supporting tapes or cords. It is used when the subsequent binding is to have a flexible cover and any tapes or cords used in the sewing might show in the binding.

Probably the most widely used hand-sewing technique is that of tape sewing. Two, three or more tapes of 5, 10 or 15mm width are evenly distributed across the spine. The path of the needle (fig. 4.8) ensures that threads are firmly and neatly supporting the tapes in position and that the 'kettle' (joining) stitch is evenly tensioned. An important feature of this method is that while the tape is held firmly against the book it is in fact loose under the thread and this has some significance in the construction of letterpress bindings. A

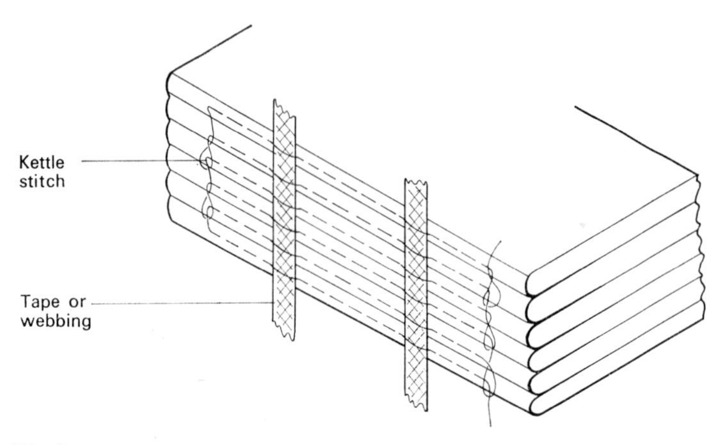

4.8 *Hand sewn on tape.*

variation for account-book bindings uses a stronger herringbone webbing instead of soft tape.

Two variations of 'cord' sewing are used when the book is subsequently to be bound in a 'laced on' style. 'Flexible' implies that the cords are left outside the book and the cover is later moulded round these to form the 'raised bands'. As this style tends to inhibit the opening of small books and those printed on stiff paper the alternative 'recessed cord' method is more often used (fig. 4.9).

4.9 *Recessed cord sewing.*

When sewing by hand the bulky thread used may result in the spine of the book being considerably thicker than the fore-edge. This is referred to as 'swell' and can be the cause of some difficulty in later binding processes. To

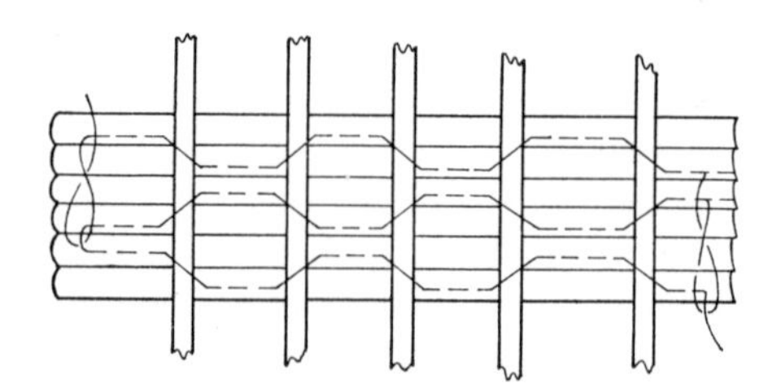

4.10 *Sewing two sections on.*

overcome this problem sewing 'two sections on' (fig. 4.10) may be incorporated at any point in the sewing of the book using any of the methods already mentioned; some loss of strength is inevitable.

The problem of sewing a book composed of single leaves and therefore no back fold is dealt with by 'overcasting' or 'whipstitching'. A clean back is given to the book by guillotining and this is then given a coat of glue. When dry, the leaves are broken off into suitable sections and using a thin thread the leaves are overcast together. When the whole book has been treated in this way the book may be sewn on tapes or cords as required.

A more economical method of achieving the same end is to overcast directly on to tapes. This is known as 'oversewing' and is often used in

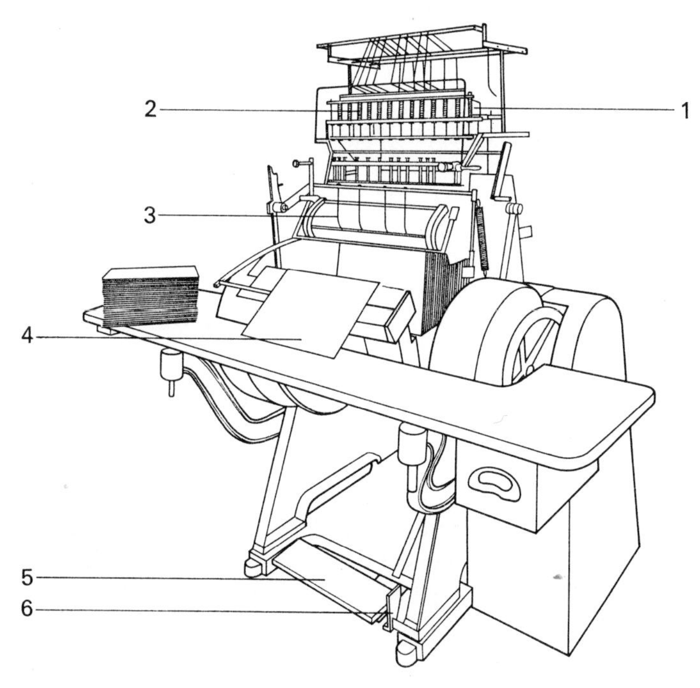

4.11 *Handfed sewing machine.*

1 *Cop of sewing thread*
2 *Thread tension system*
3 *Holdback fingers*
4 *Section on saddle*
5 *Operating treadle*
6 *Safety lock*

commercial binding of single leaves, *eg* letters, invoices, statements, etc, for filing purposes.

Machine sewing

Modern machines using straight needles may be grouped into hand-fed (fig. 4.11), semi-automatic and automatic versions; some of these are miscellaneous in character and can be set to produce several types of sewing while others can be set for only the widely used french sewing.

Unlike hand sewing, more than one thread is used and the sewing units are spread across the spine as required, one thread to each unit. The items needed on the machine to sew one unit are a sewing needle and a hook needle mounted in the needle bar, two pierces and a thread carrier mounted under the saddle; sewing thread running off a cop and through a tensioning system is threaded into the eye of the sewing needle (fig. 4.12).

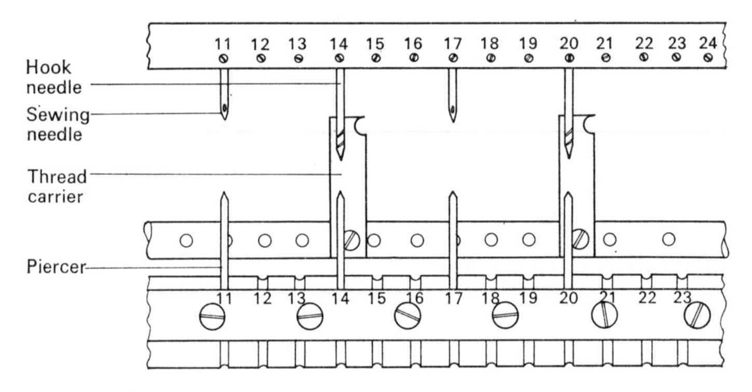

4.12 *Units in machine sewing.*

The section is laid on the saddle and up to a gauge; the saddle rises until the fold of the section is lying under the hook and sewing needle. The piercer bar moves upwards forcing the piercers through the openings in the apex of the saddle and through the back fold of the section. As the piercers withdraw, the hook and needle, carrying the loop of thread, move down through the holes and into the saddle. At the lowest point of movement the needles pause whilst the point of the thread carrier, working from left to right, engages the loop and carries it across to the hook (fig. 4.13). During the downward movement of the hook a rotational movement of 180° has taken place so that the carrier can now place the loop on to the hook of the hook needle and withdraw (fig. 4.14). At this point the needle bar rises upwards to its neutral position and the hook needle reverses (fig. 4.15). This leaves the section

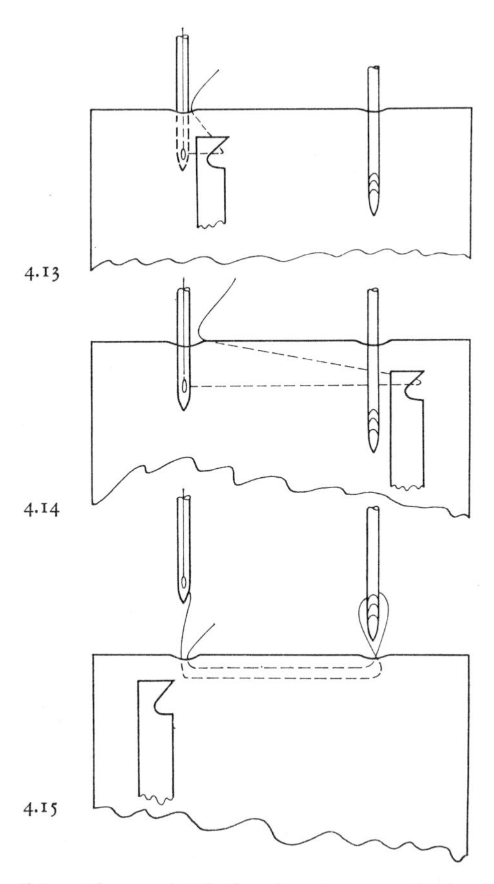

4.13 *With needle bar at lowest point the thread carrier engages the thread loop.*
4.14 *The hook needle about to engage the loop.*
4.15 *Stitch complete with twisted loop on hook needle.*

standing on the delivery table of the machine supported by a loop of thread.

During the sewing of the second section the loop on the hook needle is cast off in such a way that a chain stitch is formed. This and the single thread at the other end of the stitch firmly link the sections together (fig. 4.16). After each operation of the machine the sewn section is pushed along the table leaving the operational area clear for the next section.

A spring-tensioning system controls the thread and adjustments are made to allow for variations in section thickness and for stretching or contraction of thread caused by humidity changes. Books are sewn in continuous line

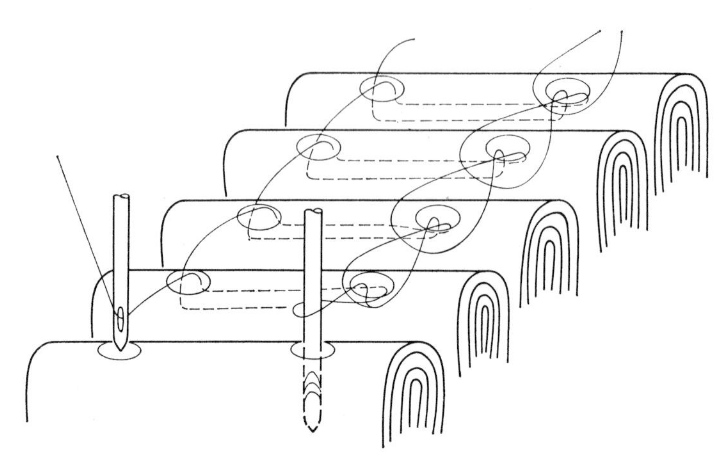

4.16 *Path of the thread in machine sewing.*

separated by a blind stitch; this is achieved by operating the saddle once, between books, without covering the saddle with a section. The books are finally cut apart by severing the blind stitch, one end of the thread being retained in the chain stitch as a lock (fig. 4.17).

These principles are the basis on which all modern sewing machines work; thus an A5 book may be french sewn with four stitch units of 30mm length and have 20mm between stitches (fig. 4.17).

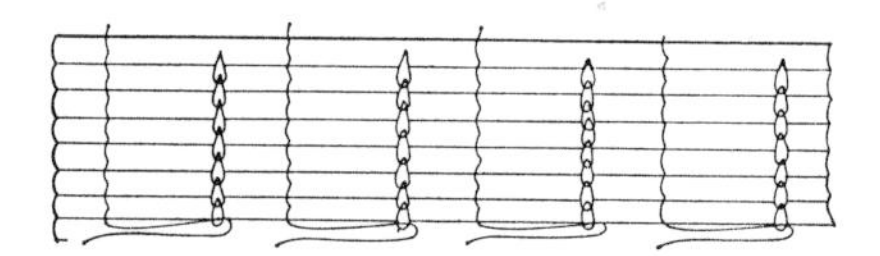

4.17 *Machine French-sewn book showing lock stitch.*

Sewing 'through' tapes implies that reels of soft tape are loaded on to the machine and fed down to the sewing position. The hook needle operates through the tape and the resultant books have the tape held close to the sections by the chain stitch. A 'tucking' device ensures that sufficient tape is allowed between books for the binding process. In a similar way a roll of mull can be mounted on the machine and all the stitches made to pass through it.

Two systems are used to produce sewing 'over' tape. One utilises a shifting sewing needle so that short and long stitches are made on alternative sections. As the needle is producing the single thread, this will now pass diagonally across the book and over the tape (fig. 4.18). The second method requires an attachment on the machine called a 'braider' with its own separate thread. The braider moves left and right on alternate sections weaving the braiding thread around the chain and single thread of adjacent units; by suitably spacing the units, tapes of 5, 10 and 15 mm may be used.

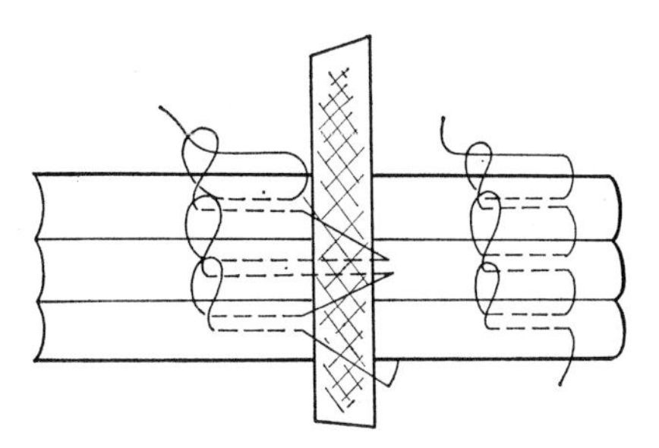

4.18 *Long and short stitches over tape.*

'Pasting' is the technique of securing the first and last two sections together with an edging of paste at the spine fold. This ensures good strength between these sections which are the ones that take most wear during the life of the binding. On hand-fed work an independent paste box contains a small feed wheel and loose pressure wheel. When the section is run between the wheels a narrow rim of paste is deposited on the underside; this is then fed on the saddle so that the paste is in contact with the appropriate section and adhesion occurs.

Although the principles remain the same, the differences between the three types of sewing machines are mainly in the method of section feeding. The hand-fed type requires that the sections be placed directly on to the saddle and this means that the operative has great control over the feeding operation but output is relatively slow. Over and through tape sewing, landscape and other unorthodox work is best carried out on this type of machine. A special setting allows the machine to accommodate books of slightly different sizes and this is very useful when quantities of single copies are to be rebound.

The semi-automatic and automatic versions are discussed in Chapter 12.

Side sewing

Like thread and wire side stitching this implies that the thread passes through the book from back to front. The gathered book will normally have a reinforced endpaper and may be wirestitched to make the book stand vertically whilst being sewn.

The machine is equipped with rotating drills as well as a vertically operating hook needle. The book is laid on to the flat working table and automatically moved forward about 20mm at a time. As the book comes under the drill units it is drilled halfway through from the underside and then halfway through from the top at two adjacent stations. At the next position the hooked needle rises through the drilled holes, engages the thread and draws down a loop. The book again moves along and the hook moves up leaving the previous loop outside the book. As the second loop is pulled down the first is cast off and a chain stitch is formed (fig. 4.19).

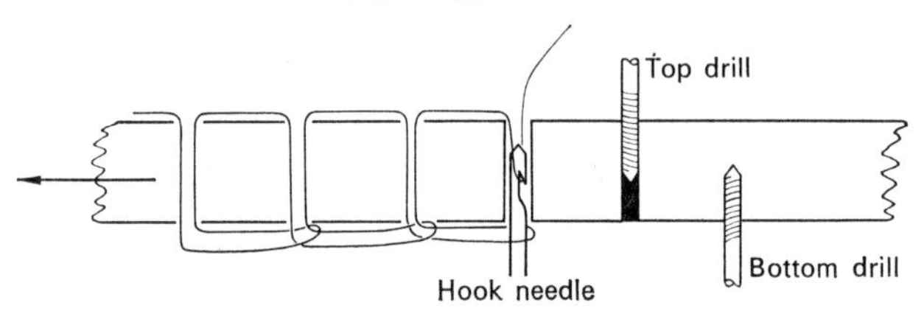

4.19 *Principle of side sewing.*

This method of fastening together the sections of a gathered book gives great strength to the bound volume but it will not open flat. It is best applied to volumes of large area and printed on thin flexible paper. Very little shape is achieved during the rounding and backing stage. The American 'oversewing' machine uses a modified sideswing principle and is specially applicable to the sewing of library rebinds. An industrial version of the domestic sewing machine may be employed for two tasks in book production. To reinforce the first and last sections of reference work (particularly bibles) the stitch line is placed about 3mm from the back fold. For single section school readers, heavier machines and threads are used to put a line of stitching along the spine fold of the inset book.

Sewing threads

Sewing threads are manufactured from linen, cotton and synthetic yarns. Linen is now only used for best hand work; cotton, nylon and terylene for most commercial applications. Threads are chosen for their tensile strength in relation to the task in hand. Thick threads will create excessive swell whilst thinner versions may fracture under the stresses imparted by the forwarding processes.

Adhesive binding

This is a method of fastening the leaves of books and magazines together as an alternative to stitching or sewing. Probably the simplest of all methods it implies that the book block is reduced to single leaves and then re-made into a homogenous unit by an application of adhesive to the spine. Today a large proportion of the square back ephemeral material published uses this technique and a wide range of machinery is available.

Four factors that need to be considered in relation to this process are (1) the stock on which the book is printed; (2) the way in which the spine folds are removed and the subsequent treatment of the cut leaves; (3) the type of adhesive used; and (4) size, speed and features of the machine.

Many of the paper stocks used for printing books, *eg* paperbacks, hardback novels, children's toybooks and annuals, timetables and guides are chosen for their bulk, cheapness and absorbent nature. These readily absorb the ink in the fast printing process employed for this class of work and these characteristics make it extremely suitable for adhesive binding. Some of the strongest bonds are obtained on mechanical printings, machine finished printings, antiques, esparto bible papers, the cheaper cream wove writings and similar stocks. These papers have an open construction, have not been heavily calendered and contain a modicum of engine sizing, all factors that allow of a deep penetration of a water-carried adhesive into the fibres with good adhesion results.

Harder papers have usually been calendered and may have furnishings that preclude successful adhesive bindings. Offet litho printings, cartridges, banks, bonds, ledgers and craft wrappings are examples of papers difficult to penetrate with water-based adhesive particularly if a starch, plastic or animal size has been used.

Coated and loaded stocks such as arts, process coated papers and supercalendered printings are heavily prepared with chalky materials. During the spine preparation process this is reduced to a fine powder and must be removed before the adhesive is applied otherwise the adhesive is deposited upon the powder instead of the paper fibres and a constructional weakness ensues.

Books may be printed having the grain parallel with or at right angles to the spine and this too may influence penetration. In the first instance relatively few end fibres are presented to the adhesive and this may give a bond of lesser strength. In the second case most of the fibre ends are in the adhesive film which may easily penetrate in and between the fibres and this may give up to 15 per cent greater strength. Of course moisture will also swell the fibres and if the adhesive is excessively wet the cross grain book may cockle in an alarming and undesirable fashion. The ink layer of solid blocks printed into the spine sometimes skins when the adhesive bound book is opened and where two facing pages are so printed adhesion between them may be difficult.

The simplest way of reducing a book to single leaves is to cut off the spine folds in a guillotine, but this gives a square hard end to the leaves and makes both penetration into the fibres and keying difficult. If the book is fanned out before gluing, the adhesive is made to penetrate between the leaves so that in effect they are rooted in a film of adhesive. Although very successful on many types of work this method is relatively slow (fig. 4.20).

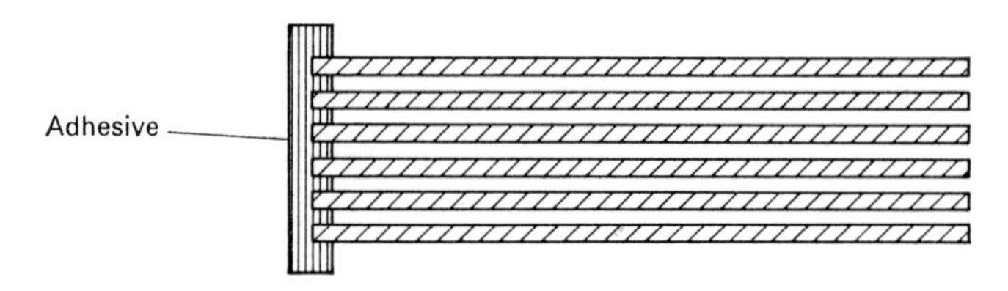

4.20 *Leaves set into glue film.*

On line-production machines plain rotary cutters, toothed saw blades and carbide tipped grinders are the usual method of reducing the book to single leaves. With the addition of other small devices such as rotating steel points, wire brushes and coarse sanders the cut edge may be further processed to produce random or patterned effects on the spine. The main purpose of these is to increase the area exposed to the glue film and thereby improve the bond (fig. 4.21).

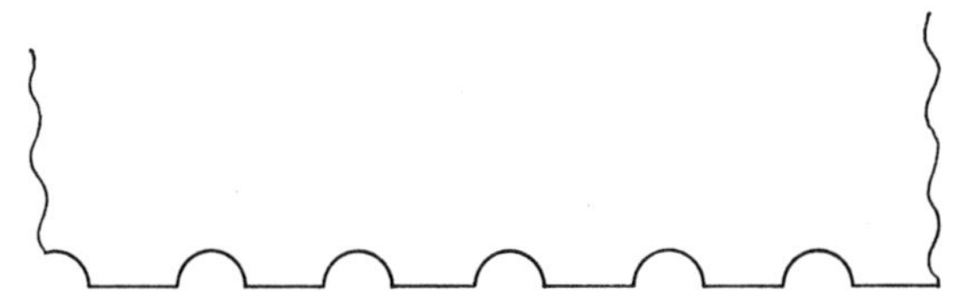

4.21 *Grooved spine showing increased surface area.*

Special grades of animal adhesives may be used for this class of work; these are used hot, are heavily plasticised and dry by a combination of heat loss and evaporation. Relatively cheap, the colour may vary from dark brown to near white (the colour having been extracted to prevent the glue line showing on the completed book). The tack of this adhesive is very good and to improve the bond a strip of cheap mull is sometimes included between the book and wrapper.

Polyvinyl acetate (PVA, is a water-based synthetic adhesive; used cold it dries as a thin transluscent film of great flexibility. The water tends to penetrate well carrying the film deep into the fibre. Early formulations had little tack and this necessitated running the machine at a relatively slow speed. Drying is also slow, two hours being considered a minimum for natural air drying. Drying mechanisms can speed the process but work is often left overnight before trimming.

The third category of adhesives used for this work is 'hot melt'. These are wax-like substances of 100 per cent solid formulation. Temperatures between 175 and 190°C are needed to bring the adhesive to working condition and sometimes pre-melting is considered necessary. Each batch is made to have a specific 'open time' suitable to the machine on which it is to be used; this is the time between the application of the adhesive and the point at which it sets off and may vary between four seconds on the small machines and nine seconds on the larger types.

Unlike previous adhesives this one dries by loss of heat; immediately the book has been glued and removed from the heat source the 'open time' scale starts to function. Within this stated period the binding must be completed because the heat loss has been sufficient to set the adhesive even though it may still be too hot to handle. Small billets of about 100 grammes weight is the most popular method of purchase; other industries use the same adhesive

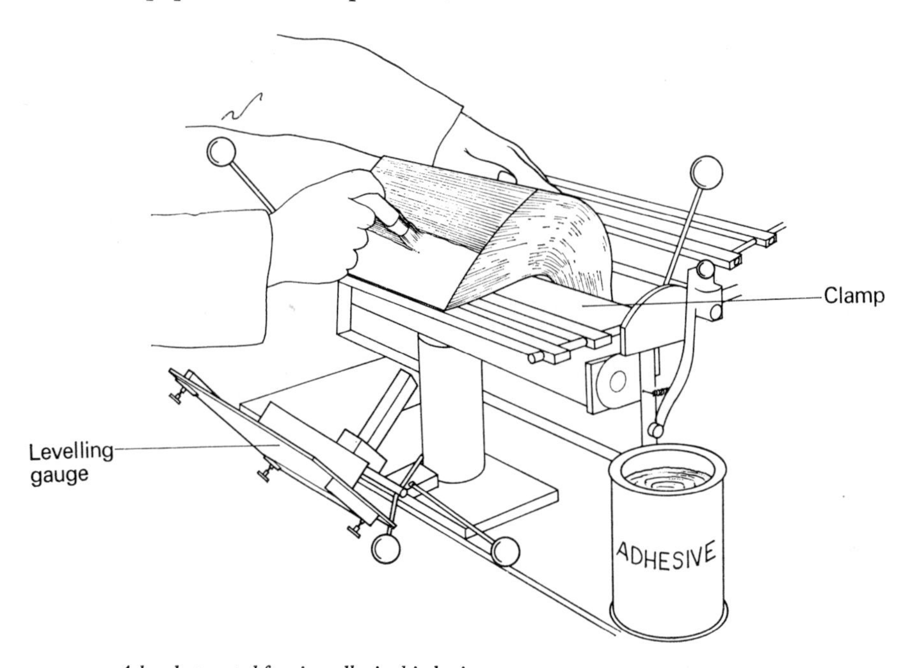

4.22 *A hand-operated fanning adhesive binder in use.*

in plug and cord form but of course these require special applicators.

Hot melts do not penetrate into the fibres as far as PVA but appear to have greater film strength and rigidity, the resultant book having a stiffer back. Used on its own (one shot) it is particularly suitable for cross-grain work and the harder range of papers. Some of the machines are equipped to apply a primer of PVA, which is dried on machine and then coated with hot melt for cover attachment. This (two shot) system is claimed to produce a better binding as it blends the advantages of both types of synthetic adhesives. The PVA penetrates the fibres and gives a strong bond while the hot melt provides the fast method of cover attachment and the stiff back for support.

Machines and devices for adhesive binding range from a simple bench-mounted device costing just a few pounds, through a considerable number of intermediate sizes to a very large 70-clamp model that may be linked into an automatic gathering-binding-trimming line.

Three principle types of end product are catered for (1) padding; (2) wrappered book or magazine; and (3) a book lined with calico, paper or cloth for subsequent flush binding or pasting into a publishers case.

The padding press is really a form of adhesive binding device and is simply a low cost press that will hold a stack of paper under firm pressure while the edge is glued and dried.

'Fanning' is an intermediate method that may be accomplished on a bench- (fig. 4.22) or floor-mounted model, the adhesive being applied by brush or roller. Much in demand for heavy books is one machine that has three radially mounted arms. At the three production points the operatives prepare the book in the clamp, apply the adhesive (PVA) by rotating against a glue roller and finally closing the spine with a pressure clamp and applying linings before removing the book from the machine.

Machines that produce wrappered books may have one or more clamps into which the book is fed, jogged and clamped under spring pressure (fig. 4.23). The spine is then carried progressively across a cutter or grinder, glue mechanism, covering station and creasing station to delivery. Production speeds may be from 100 to 9000 books per hour depending upon the size of machine, length of run, adhesive and other variable factors. Books of suitable size may be produced two-up and with special attachments the larger versions of these machines can be set to produce calico-lined books as an alternative to the normal covering style.

Adhesive-lining machines that produce the third category of product progress the work forward by means of canvas belts. The backs are milled or cut, glued and then joined to a continuous strip of glued lining material. After turning the overlap on to the sides, the strip lining if firmly rubbed into posi-

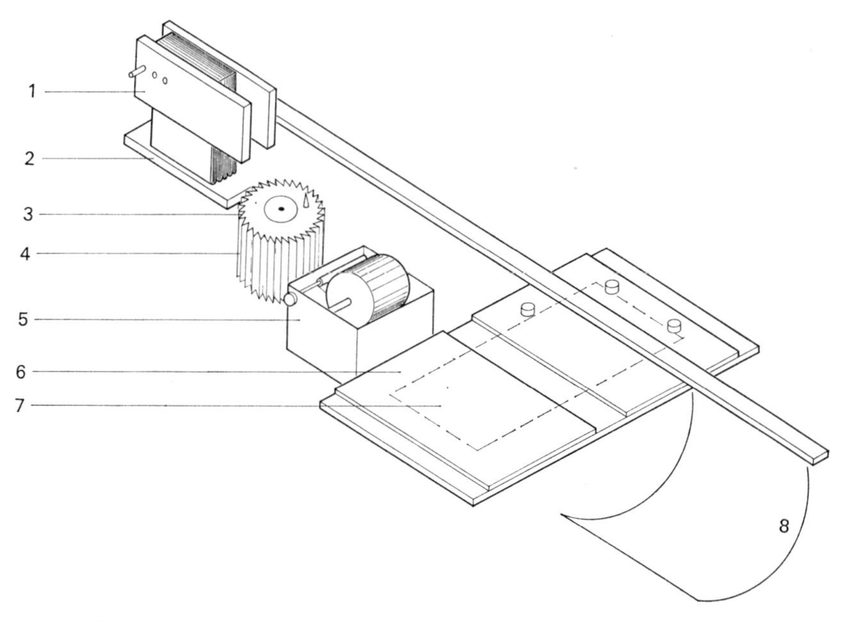

4.23 *Elements of the adhesive binding covering process.*

1 *Book clamp and head stop*
2 *Levelling plate*
3 *Cutter with inset stylus for grooving*
4 *Dust extraction fan*
5 *Glue tank*
6 *Cover creasing plates*
7 *Cover in position against feed stops*
8 *Delivery trough*

tion and the books are separated by high-speed knife (fig. 4.24). On these machines the actual production speed is relative to the length of the book being processed and the running speed of the belts. If the belts are moving at 8m per minute then 35 books of A5 size, plus 20mm between books, may be fed every minute. A larger A4 volume, including gap, will be fed at 25 books per minute. These machines are usually hand fed and equipped with variable speed mechanism to allow for jobs with different handling problems.

Both adhesive covering and adhesive-lining machines can be set to process work that has been previously wirestitched or thread sewn, the cutting head being removed or withdrawn.

Paperbacks and magazines are intended to have a relatively short life and

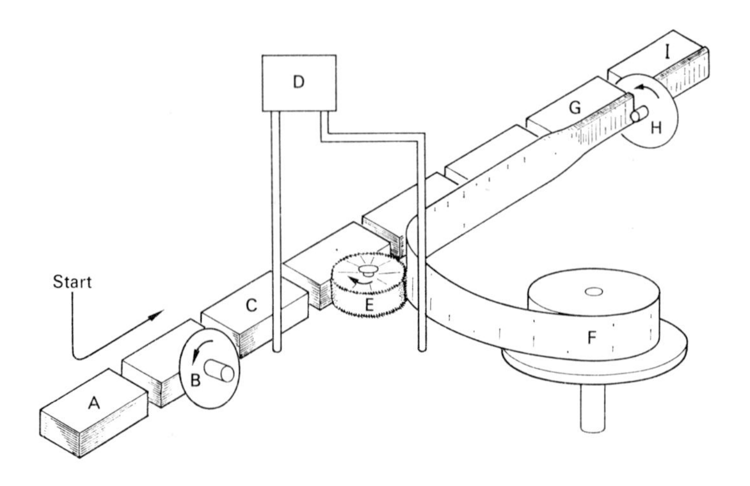

4.24.1 *The adhesive lining process.*

A *Gathered book*
B *Knife cuts off folds*
C *Book in single leaves*
D *Gravity fed glue pot*
E *Glue brushed on*
F *Expanding calico reel*
G *Calico moulded on book*
H *Separating knife*
I *Delivered book*

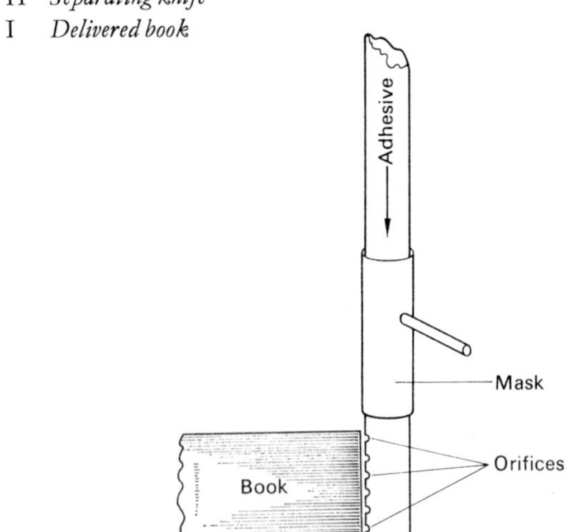

4.24.2 *Detail of the glue mechanism.*

it may be acceptable if the leaves remain firmly in the cover for only one reading of a paperback or until the next issue of the magazine. Casebooks in the cheaper range may be produced more economically by adhesive binding but at the present time publishers of quality hardback books have serious misgivings about the permanence of adhesive binding and few are prepared to have their work bound in this way.

5. Simple binding methods

In this chapter the combination of some of the processes mentioned in earlier chapters into the better known styles for booklets and magazines are discussed.

Inset work

The sequence of production for inset (or saddleback) work is insetting, stitching and trimming. Sections placed one inside the other are said to be 'inset' and this is a basic form of construction for work that has to be produced cheaply or has a small number of pages. The units may be of different thickness, colour or type of paper but ideally the plan should be to place the heaviest section in the centre of the completed booklet. In this way jogging will bring the heads down to register most easily; a 60-page booklet may be planned for assembly as cover, 4pp, 8pp, 16pp, 32pp in centre.

The fastening by wirestitching is completed by two or three stitches passing through the back fold. Stitches may be placed or staggered; the former guarantees accuracy of position whilst the latter distributes the extra thickness of the wire so that a pile of stitched booklets will not swell unnecessarily. Swell is a contributory factor to inaccurate trimming. Best work may be thread stitched.

A useful attachment on the folding machine combines folding and fastening. This is achieved by the application of a thin line of adhesive at a critical point during folding and when trimmed the leaves are seen to be glued together. There is of course a limit to the number of leaves that can be so treated.

Most inset work is trimmed 'flush', *ie* text and cover trimmed to the same dimensions, and are handled in piles of suitable height on a single-knife guillotine or three-knife trimmer. Overhang covers are sometimes required and this can be done by (1) insetting without cover; (2) wirestitching with one security stitch in the centre of the spine fold; (3) trimming text to size, trimming and folding cover to have suitable overlap; (4) insetting text into cover; and (5) wirestitching with two wires ensuring that the text is accurately positioned inside the cover. Variations of this style include the use of fancy cords and tassles for presentation work; covers turned in on the fore-edge and pockets wired into the back and front covers to receive inserts. The term 'insert' implies the addition of material to a book or magazine after it is bound. The production of large quantities of saddleback work is speeded up by the use of in-line units and these may include automatically fed insetting and wirestitching, single copy trimming, inserting, and mailing mechanisms.

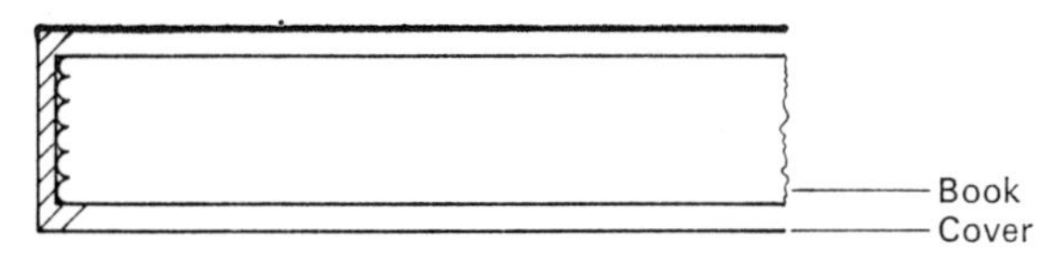

5.1 *Wrappered book.*

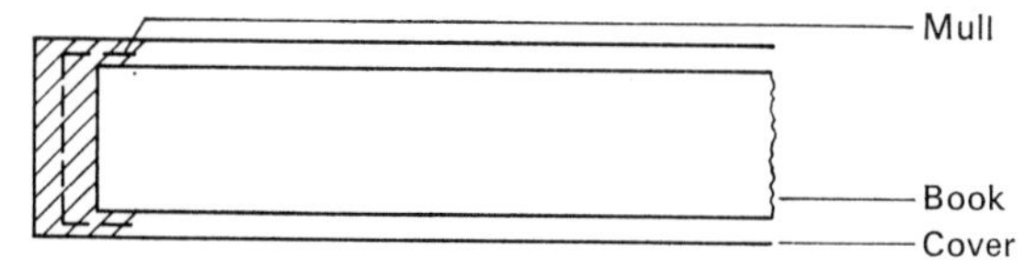

5.2 *Wrappered book with mull lining.*

Wrappering

This is widely used for square back magazines and paperbacks and this term implies that the cover is adhering to the spine only. The book block may be fastened together by the previously described side wire-stitched, stabbed, sewn or adhesive bound techniques.

Small quantities are readily produced by hand; the spines are glued with tacky animal glue or polyvinyl acetate and the cover drawn tightly round the book block. It is important that the cover material is in good contact with the spine while the adhesive dries and care taken in trimming to retain the square professional finish. To help the accuracy of pitching the cover it may be necessary to print a small pitchmark between pages three and four (fig. 5.1). Thick varieties of cover stock will require creasing beforehand.

'Covering' machines have six or eight clamps and work in a circular form. The book is dropped into an open clamp which grips and passes it over the glue mechanism. The cover is dabbed up from a pile and adheres to the tacky spine, delivering after suitable creasing and rubbing down stations. Trimming follows as a separate operation. For economy reasons these machines are usually charged with hot or cold animal glue with good tack qualities. These machines have now been largely superseded by the multi-duty versions that combine wirestitching and adhesive binding with the wrappering process.

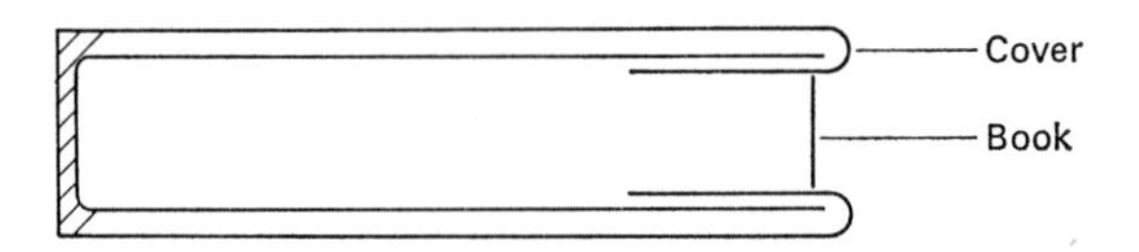

5.3 *Wrappered book with overhang and tuck foredge cover.*

As with saddleback work variations can include overhang covers and wallet fold fore-edges but in each case the trimming operation must be re-positioned in the sequence of production (fig. 5.3). If the book block is side wired the wires may be covered by double creasing the cover and ensuring that some of the adhesive flows on to the face of the first and last sheets of the spine (fig. 5.4).

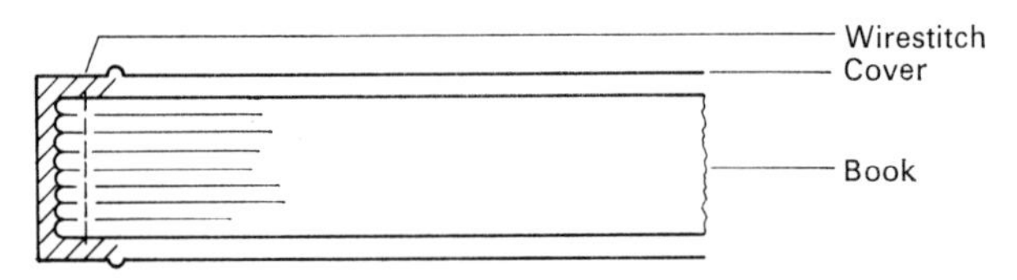

5.4 *Wrappered book wire stitched and side glued using a creased cover.*

A variation of wrappered work that is much used by publishers of children's primers, has the cover glued to the end leaves of the book as well as to the spine. Usually described as 'cover drawn on' it is essential here to incorporate blank leaves in the book or to tip on endpapers (fig. 5.5). Actual covering is a hand operation although means of mechanising are adopted when quantities are sufficient; this includes the use of sheet gluing machines described in Chapter 10.

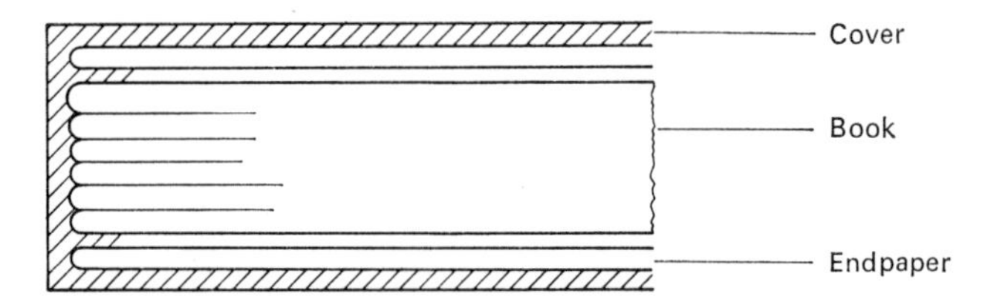

5.5 *Cover drawn on.*

Mechanical binding

This term has been adopted to describe the many styles of fastening single sheets together by inserting shaped metal or plastic units into holes punched

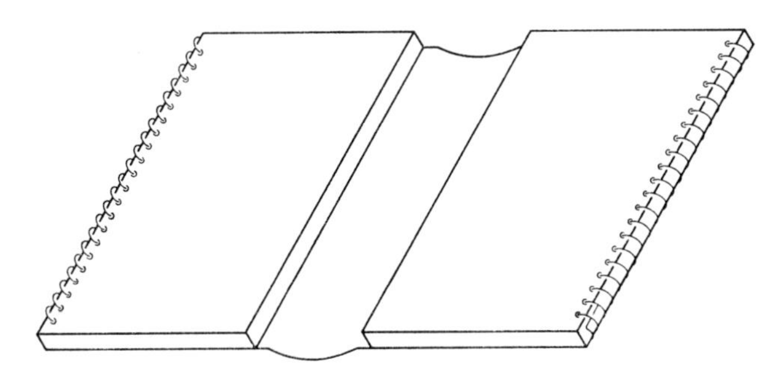

5.6.1 *Spiral foredge style.*

into the sheets. Normally sheets cannot be removed or added and these methods should not be confused with true loose leaf mechanisms which are discussed later in the chapter. Three popular styles are metal spiral or coil, 'Wiro' and plastic comb.

Spiral binding machines usually accept the punched sheets, form a spiral coil from a wire spool, insert the coil into the holes by rotation, cut off the correct length and then for safety reasons turn-in the ends into the coil.

'Wiro' is a preformed wire comb construction of various diameters. The correct length is cut and inserted into the round holes or slots that have been previously punched into the sheets. A closing press is used to press the semi-circular wire into a complete circle and the binding is complete.

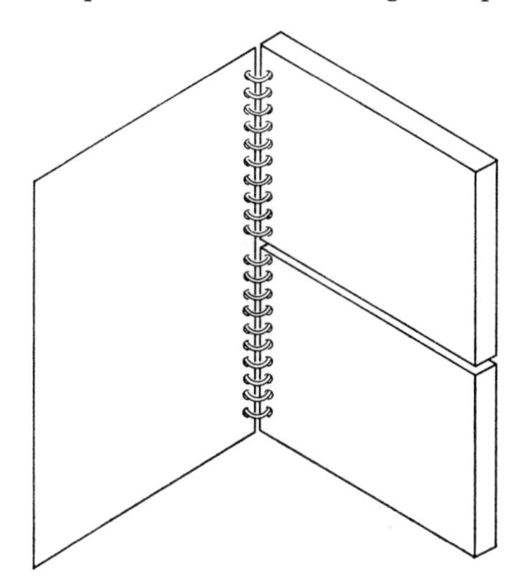

5.6.2 *Split page work 'Wiro' bound.*

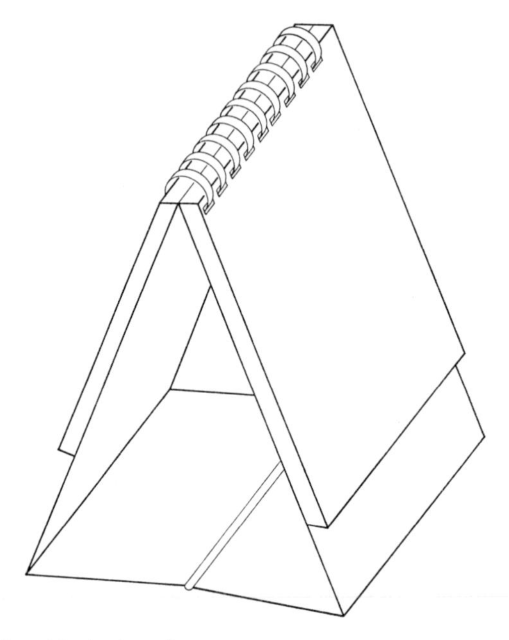

5.6.3 *Easel binder with plastic comb.*

Plastic combs are stamped out of suitable material and while still in the flat condition may be foil blocked with title or decoration. These are then heat formed into a circular shape of given diameter and allowed to cool. Binding involves mounting the comb on to a device which gently pulls open the prongs allowing the sheets to be fed on to them; the comb is then released and the binding is complete.

The coil action of spiral binding gives a vertical action to the leaves when the book is opened and makes this style unsuitable for work that has horizontal printing alignment. Jobs that function by turning the leaves through 180°, *eg* reporters' notebooks and calendars, may be bound on all three methods, the plastic comb having a special narrow back comb for the purpose. Other specialities include split page work, easel binding, fore-edge openings, and wallet construction (fig. 5.6).

Loose leaf mechanisms

The production of loose leaf mechanisms may form part of the normal stock in trade of a miscellaneous commercial and stationery bookbinder, but it is frequently the subject of intense specialisation by very large companies. In the latter case many of the devices marketed are heavily protected by patents and it is impossible to list or define the very wide range offered. There are however certain basic forms that can be described that are utilised by both the specialist and the miscellaneous houses. To qualify as a true loose leaf mechanism it must be possible to add and remove sheets at will, the binder offering good security to the leaves when closed.

The customer's order for a loose leaf binder will specify the type of mechanism required, capacity of the spine, page size, ruling and printing, hole positions, whether index sheets and endpapers are to be supplied, cover material and colour, external titling required. To ensure an accurate match a specimen leaf is often lifted from an existing binder.

Interscrew

These are brass, aluminium or plastic rods of about 6mm diameter with a fixed head at one end and screw head at the other (fig. 5.7.1). Available in lengths from 12mm upward to suit the job, they are usually used in conjunction with some variation of the hinged board technique (fig. 5.7.2).

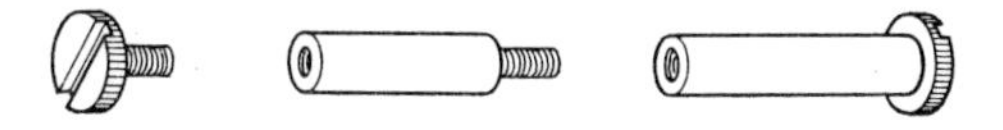

5.7.1. *Interscrew and extension.*

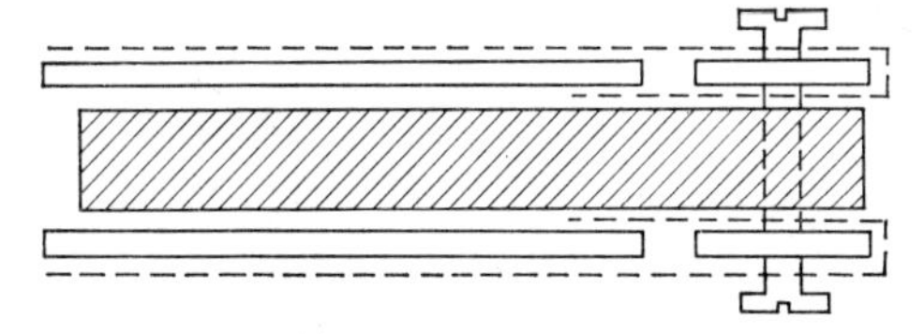

5.7.2 *Interscrew used in conjunction with hinged boards.*

Ring metal

A number of metal rings are fixed to a metal backbone; these are either complete rings or are split in the centre and hinge at the attached end. In cross section the rings may be round oval or flat and in diameter range from 12 to 50mm. The number of rings used controls the ultimate cost of the binder; cheap binders often have just two rings 80mm apart but these give little control to the sheet which readily moves and tears (fig. 5.8.1). The most expensive in the range are the multi-ring metals (both split and solid ring) with rings as close as 12mm centres. These give excellent control of the sheets and make for maximum wear (fig. 5.8.2).

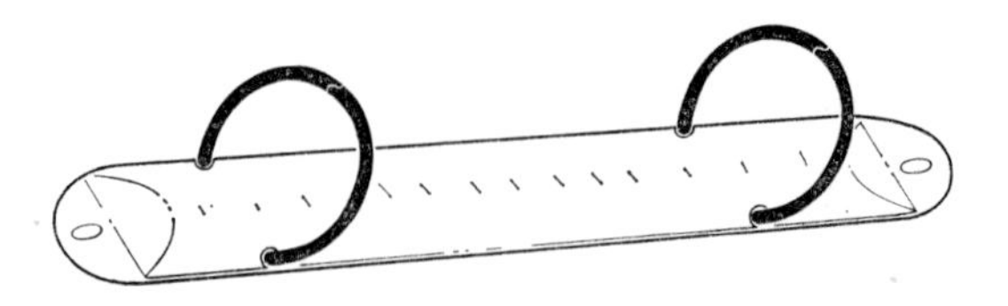

5.8.1 *Two-ring flat back metal.*

Ring metals are fastened to the spine section of a three piece case and in the cheaper styles the metal is riveted to the board hollow giving a flat back to the finished binder. To give a more bookish appearance to the better quality binder a convex spine is desirable; this is achieved by a shaped metal plate being inserted into the cover construction and to this backbone the ring metal is affixed.

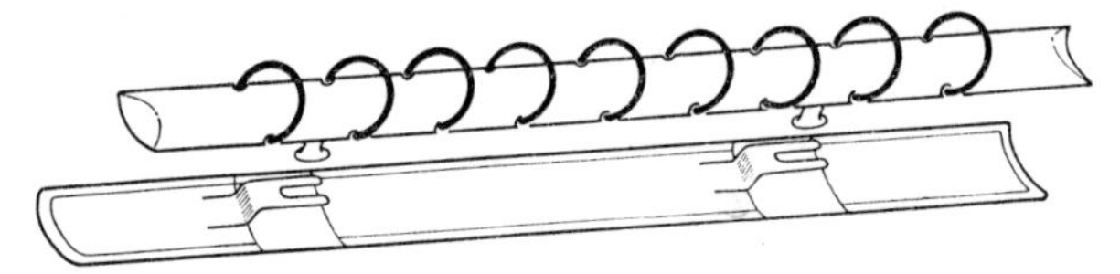

5.8.2 *Multi-ring metal with convex metal spine attachment.*

Leather, leathercloth, fabrics, papers and plastic cover materials are all used in the production of ring binders and many standard sizes are also available

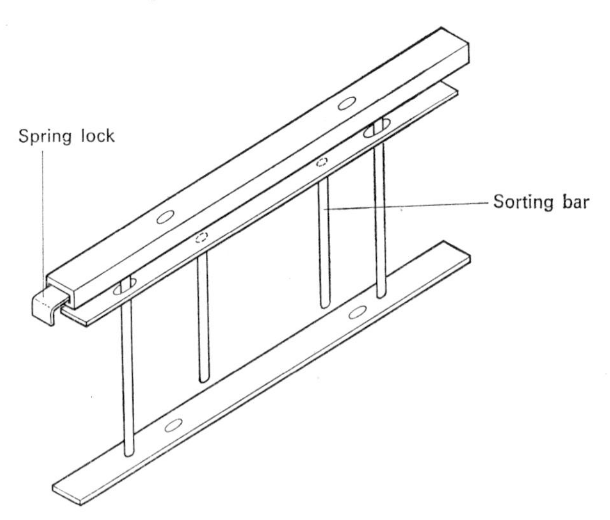

5.9.1 *Universal metal.*

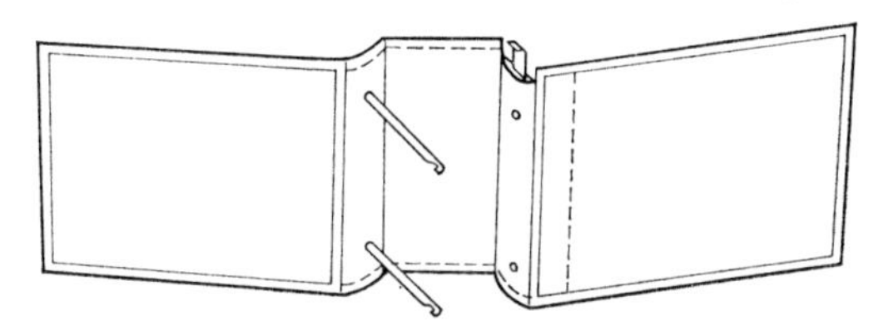

5.9.2 *Universal binder.*

in high frequency welded plastic cases. One very cheap variety features a single piece of strawboard suitably covered and then creased to provide the hinging points of correct spine width.

Record or universal metal

This is another mechanism that is affixed, into a five piece case, by the bookbinder (fig. 5.9). The straight post metal is riveted into the narrow flanges of the case so that when closed the posts clip into the spring operated locking box; a two-posted sorting device is supplied with each metal.

Strong fabric cover materials are used for the cover and a lining over the hinge usually covers the metallic parts. As a reasonably priced mechanism it is frequently used for the storage of completed pro-forma.

Endlock

This metal is used in conjunction with a hinged board construction, the bottom board containing anchor posts which are screw cut to accommodate extension posts. The top half of the mechanism engages on the posts and an end key rotates a screwbar to friction lock the top assembly on to the posts (fig. 5.10).

Prong metal

A mechanism has curved prongs which enter the paper block from both sides; it is constructed to open partially so that the paper does not become disengaged from the prongs. This allows it to be used bookwise, the pages lying flat and making it useful for the keeping of records that require updating by hand (fig. 5.11).

Thong metal

This is in effect two strong boards linked together with flexible fabric thongs on to which the paper is looped. The boards are of hollow wooden/metal construction and the nylon thongs are anchored into one board. The thongs are

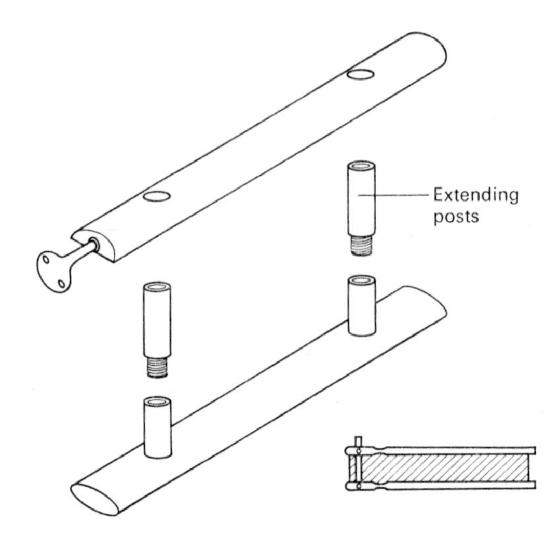

5.10 *Endlock metal and binder.*

passed through the metal clamping bars and then into the other board. Here they are anchored to a tension bar that is attached to a screw thread. A key passed into a hole in the fore-edge of this board will locate with the screw bar and when turned will allow the tension bar and the thongs to move in or out, thus expanding or contracting the capacity of the binder (fig. 5.12).

A very robust mechanism, its greatest asset is the good control achieved on

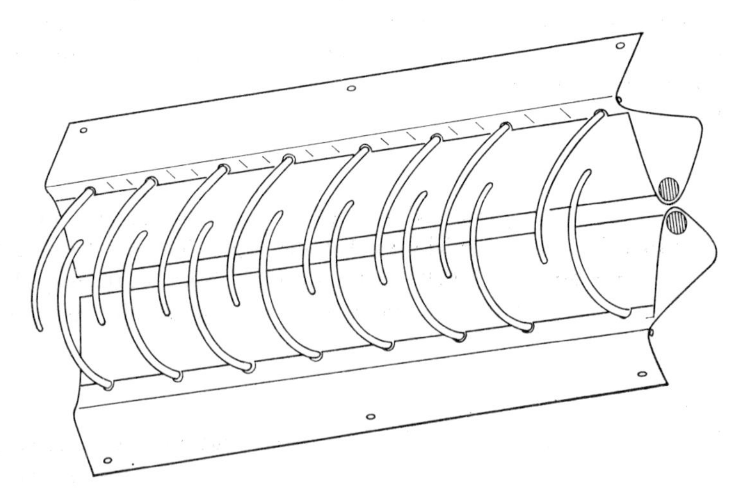

5.11 *Prong metal.*

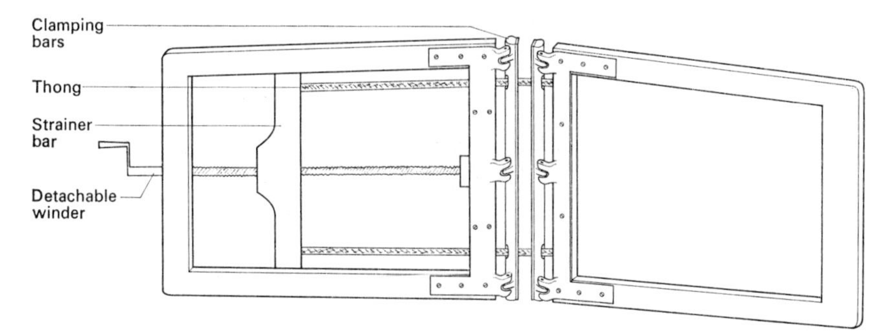

5.12 *Thong metal with board linings removed to show mechanism.*

the sheets by the wide thongs. The binder can be closed to clamp a single sheet or may be opened to contain up to 120mm of paper.

Metal back ledger

A very heavily constructed mechanism has round telescopic posts to which the paper is affixed; a key placed in the end of the spine unit engages a screw bar to which are attached a number of hinged levers. The other end of these levers are hinged to the clamping bars and when the screw thread is rotated by the key the levers move along the screw and force the clamping bars to move in or out. As the telescopic posts are attached to the clamping bars these also increase or decrease in length (fig. 5.13). The metal spine of the mechanism does not alter in size and its width covers the clamping bars when these are at its minimum capacity of about 50mm; maximum capacity is in the region of 110mm of paper. The mechanism is purchased complete with mechanical spine, boards, hinge pins and key; certain parts are nickel plated and these are left uncovered in the completed binder.

Covering loose leaf mechanisms

Very heavy quality materials are necessary to give the binder a long life and the task of fixing these around awkward shapes and to metallic surfaces, summarises the principle problems of covering loose leaf mechanisms. Pigskin, sheepskin, hides, heavy leathercloth, soft sailcloth and corduroy are the types of material used. These require very tacky adhesives and in some instances will have to be of a variety that will readily adhere to metal. A well tried technique for ensuring adhesion is to clean the metal of grease, lightly etch with oxalic acid, warm the metal slightly and then cover with soft brown paper using tacky animal glue. Once lined in this fashion, covering proceeds,

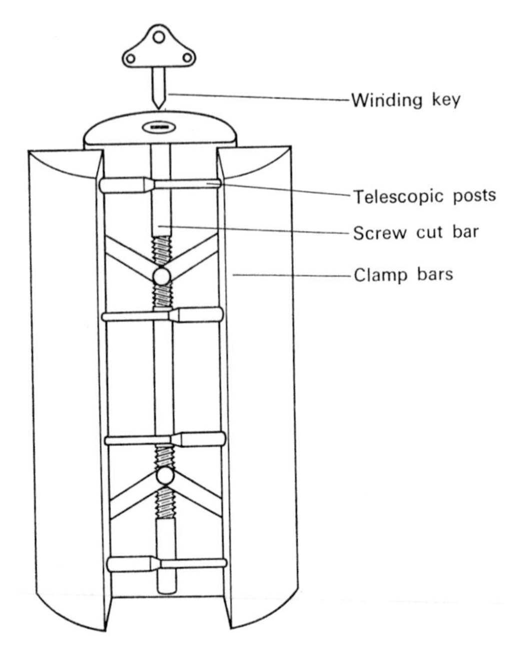

5.13 *Mechanism of metal back ledger.*

the cover material readily adhering to the lining paper. Many commercially produced binders are assemblies of metal, plywood, cast plastic and rubber shapes. These have little connection with the bound versions and engineering techniques are used in their assembly.

Furnishing the binder

To be of value to the customer the covered mechanism will require to be furnished with paper of the desired ruling and perhaps end papers, index and dividing boards. The paper may be ruled, round cornered, creased and sometimes numbered to suit the circumstances. These processes are dealt with in the relevant chapters. The index fitted is usually of the tab type, individual tabs being produced in leather, cloth or plastic and attached to the appropriate sheet or divider board.

Heavy quality, cloth reinforced, endsheets may be fitted to the more expensive types of binder and these protect the first and last sheets of the ruled stock. Dividing boards may be similarly constructed but may also be of sheet plastic with tab index incorporated. Security may be obtained by building a key-operated lock into the mechanism when it is made.

6. Binding techniques and principles

Hand binding is concerned with the preparation of suitable materials and their assembly into units of different constructional characteristics. Each style may be said to be suitable for a specific purpose and, within limits, be completed to a price to suit the customer. By varying the materials used, wide differences of price can be achieved. By definition a 'binding' is a construction that is built up around the book block, covered and then finished out. A casing has the book block and the cover produced as separate units which are then joined at the final stage.

Printed books are described as 'letterpress'; this rather narrow and archaic description has been left over from the early days of printing when all books were printed by this process. This group includes casebooks, yapp and limp work, library rebinds, albums and cord sewn work. Books for writing in are known by the generic term 'stationery' and includes flush bindings, account books, loose leaf work and guard books.

Although productive processes are similar in both fields they tend to be separated in different binderies. This is because the aesthetic and constructional considerations are so dissimilar. The emphasis in stationery binding is on strength and lasting qualities; thick heavy materials and robust constructional methods are the criteria. Letterpress books are often smaller and, being hand held in use, need to be of lighter, neater construction and may even be required for display purposes.

The early stages in production have been dealt with under the headings of folding, make-up, gathering and securing operations. The next group of operations is known collectively as forwarding and covering.

Endpapers

The function of the endpaper is to link the book to its cover, to hide the constructional features of the covering and to protect the first pages of the text. As the endpaper is often the first part of the book inside the cover to be seen, an attractive and decorative feature may be desired. As a general rule the most complicated book construction carries the most complex endpaper.

Good-quality stock, often chosen to match the text paper, should always be used for the construction of endpapers. Suitable papers are cartridges, ledgers, good quality cream woves and bonds, mould and hand made papers, the better range of cover stocks and some printers' board substances. Lining-in paper is usually thin and decorative and does not help the strength of the endpaper. Dimensional stability and good flexing characteristics are essential. Grain should run parallel to the spine to avoid cockling when being attached and when being pasted down, and to exert the correct pull on the cover when attached to it.

The simplest endpaper is the utilisation of the first and last leaves of the book. In printed books these may be left plain or be printed, but often the text stock is most unsuitable as endpapers. It may be a soft, lightly sized printing paper and this will give considerable trouble during forwarding, particularly at the pasting down stage; it has poor dimensional stability, low tensile strength and flexing characteristics. Books printed on offset printings and cartridges may have acceptable self-endpapers providing the holes made by sewing are not considered objectionable.

Single-leaf endpapers may be used for side-wired books and these are usually reinforced along the spine edge with strong paper or fabric. This ensures that the endpaper will not tear out when the book is in use.

A folded 4-page tipped on is widely used for case-books and similar styles. Attached before or after sewing by a 3mm edging of paste or PVA, this endpaper is the minimum standard for books opening right into the spine. For certain heavy books a 20mm strip of calico is pasted and drawn round to be 5mm on the section and 15mm on the endpaper (fig. 6.1).

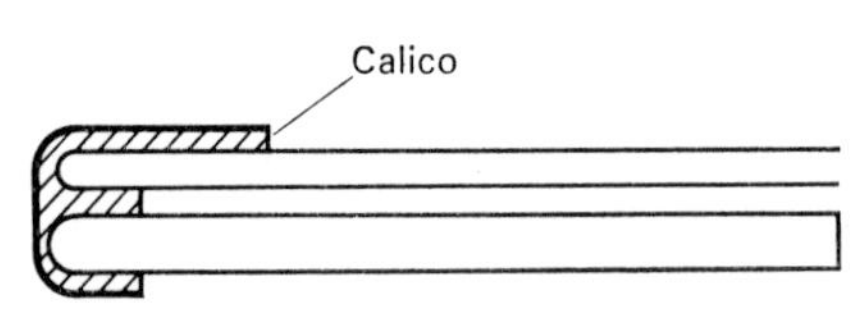

6.1 *Endpaper tipped on and reinforced.*

A strong method of attachment for 4-page endpapers that is often used in simple stationery work is to coat the outside of the endpaper with adhesive and attach it to the first (or second) leaf of the book section. This provides a strong 'stiff leaf' at each end of the volume (fig. 6.2).

A construction used in traditional account books and for some library rebinding requires two folded 4-page plain sheets, two lining-in sheets and a strip of cloth or buckram. Manufacture starts by joining the two 4-pages with the strip of cloth; the linings are then coated with adhesive and laid on the insides to give the necessary finish (fig. 6.3). This type of endpaper is usually sewn on and is known by the stationery binder as a 'joint'.

Two folded whites and one folded colour are needed for a typical 'extra letterpress' endpaper. A folded white is pasted all over and attached to the colour; after pressing and drying this unit is tipped into the remaining folded white. The completed endpaper may be tipped or sewn into position on the book. Variations of this method allow cloth or leather hinges to be added to the opening.

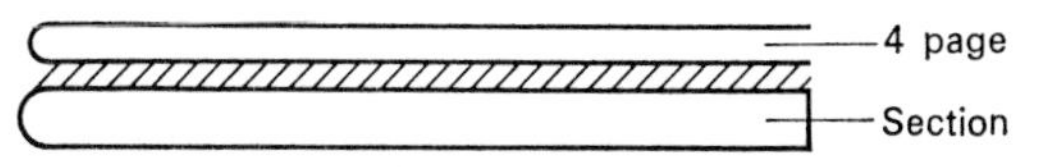

6.2 *Stiff leaved endpaper.*

Endpapers are positioned accurately to be level with the spine and heads of the section; failure to do this will introduce problem areas and malfunctioning at the hinge of the completed book.

4-page endpapers tipped to sections that are to be adhesive bound may be tipped away from the spine by the depth of the cut that will reduce the book to single leaves. This will ensure that the completed book will have the

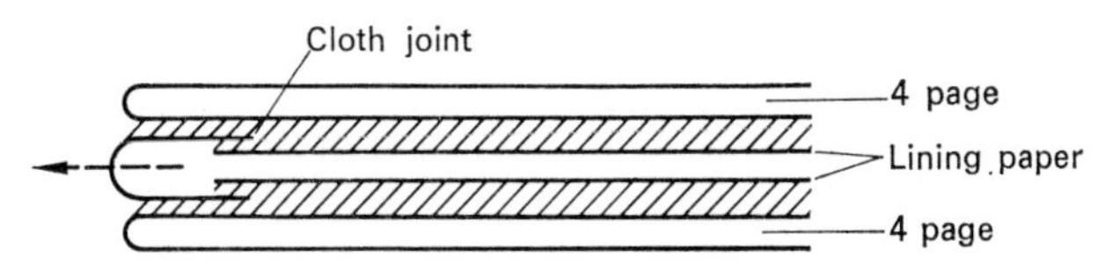

6.3 *Cloth joints for account book and library styles.*

advantage of a fold at the hinging point of the endpaper rather than a visible line of glue. Many variations of the endpapers described are used; some are more complicated but usually the simpler versions are chosen to save both materials and time.

Reducing swell

Swell is the term used to describe the extra thickness gained by the book as a direct result of the sewing operation. If a thirty section book is sewn with thread of ·25mm caliper then a swell of 7·5mm has been introduced. Methods of controlling swell in the sewing process are discussed in Chapter 4, but it is nearly always necessary to take some action during forwarding to consolidate the spine and to reduce the swell to the desired state. Swell in relation to book shape is discussed in the section dealing with rounding and backing.

In hand work the swell is beaten with a hammer until it is judged to be reduced sufficiently. In effect the paper compresses about the thread and therefore the success of the operation is in direct proportion to the hardness of the paper. A book of many sections on thin hard paper is notoriously difficult to control and steps must be taken at the sewing stage to prevent the problem getting out of hand. To deal with quantity work mechanical presses are used.

Spine gluing

Examination of a sewn book will show that while the sections are linked together with thread, the book block is in no way a homogenous whole. This is achieved by an application of adhesive that is carefully brushed between the sections so that a minimal film is visible after drying.

The spine folds of letterpress books tend to be rounded and books in this category require a thin flexible adhesive preferably of low moisture content. In practice this means a high-grade flexible animal glue or a suitable polyvinyl acetate with minimum water added; excessively wet adhesives may swell the spine unnecessarily. For gluing the spine of account books a thicker adhesive is needed to fill the space caused by thick sections and the sharp fold used.

Drying may take from ten minutes to two hours depending upon the type of adhesive used and the atmospheric conditions pertaining at the time. Distortion that occurs at this stage is very difficult to eradicate and great care should be taken to ensure that the book is allowed to dry undisturbed with the spine at 90° to the endpapers.

Book trimming

A trim in finishing terms is the removal of 3mm of paper from the fore-edge, tail and head in that order; this reduces the book to the required size and gives clean edges in the completed book. Larger trims are sometimes planned especially for bled work; in re-bound books the guillotine operator may take only the smallest possible trim in an effort to preserve margins.

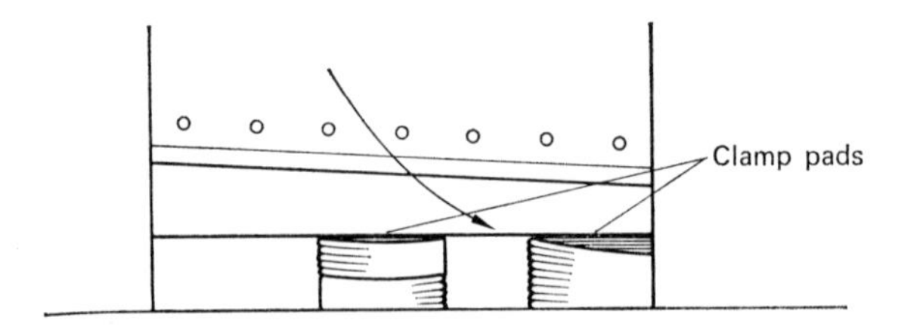

6.4 *Cutting of single and multiple books.*

At this stage the book has an irregular shape with swell distortion at the spine. When laid on the guillotine ready for the fore-edge trim the uppermost leaf will be trimmed to a slightly longer dimension than the bottom leaf. In hand binding this small error is accommodated in the positioning of the boards but in mass production work this factor may have a significant effect upon subsequent operations.

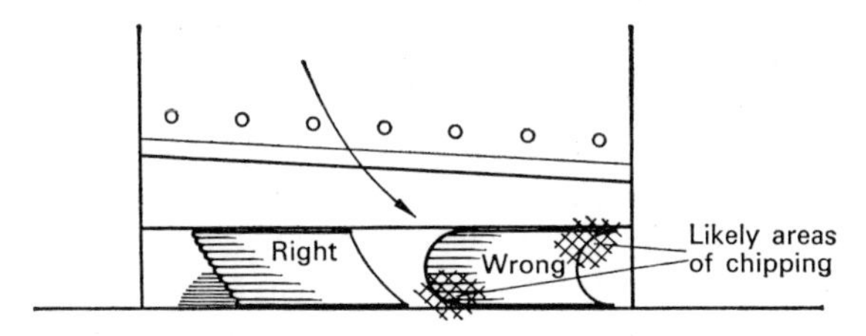

6.5 *Cutting of rounded books.*

As with other trimming operations good clamping is essential and when trimming head and tail a good 'pad' is necessary to accommodate the swell (fig. 6.4). The fore-edge of a single book is placed against the side plates of the machine so that the knife cuts into the folds of the sections first (fig. 6.4). Two or more books are reversed to distribute the swell and a pad of a different shape is used (fig. 6.4). Very poor quality paper may chip if the work is not securely clamped or if the above rule is not observed. Books with rounded backs are a special problem; chipping at the shaded areas (fig. 6.5) may be avoided if the book is reshaped and the vulnerable areas supported.

Edge decoration

Book edges may be treated to make them more attractive or to provide a protective coating. Gilding of quality books gives a rich and luxurious appearance by the application of gold leaf; this also acts as a protection against attack by air-borne injurious acids in industrial atmospheres.

To gild a book in the traditional way the book is clamped and the edge scraped smooth with a steel scraper; the open fibres of the edge are then filled with a paste made from earth pigments and brushed smooth. A thin solution of egg albumen is brushed over the edge and immediately the gold leaf is laid on to the wet surface. After drying for about ten minutes the gold is impressed into the book edge with a hard agate stone burnishing tool working through paper. When the edge is thoroughly dry it is lightly waxed and the gold burnished with the burnishing tool in direct contact with the gold.

'Plain' gilding, with the book in the square shape, is the usual commercial procedure. Gilding after rounding and backing is described as 'gilt solid' and this may be further embellished or 'gauffered' by impressing heated brass tools into the edge. For certain religious work the edges are washed over with aniline dye before gilding commences.

In an alternative dry process that has been developed for quantity work the books are clamped in a machine. After the edges have been smoothed by the application of mechanical rotary sanders and surplus dust vacuumed off, a thin coating of shellac varnish is applied and allowed to dry. A specially prepared foil is laid on the edge and heated rubber rollers, working through the foil, reactivate the varnish making it sticky. The release wax of the foil melts away leaving the gold film attached to the varnish preparation and the book edge. The foil carrier is then taken off and the books removed from the press of the machine. The edge is gently fanned to break up the film. Various coloured aluminium foils may be substituted for genuine gold; coloured edges under gold are worked in the same way as the wet process. Round corners are usually gilded individually before the flat edges are commenced.

To colour the edges of books, aniline dye is washed over the edge and a more or less permanent stain results; small quantities are completed by hand using a sponge. A wide range of colours is available in powder form which is mixed with water in the strength required.

Larger quantities of books may be sprayed using an air spray gun. The work is carried out in a spray booth where the books are stood on a raised table and a weight placed on top. During spraying the jet is directed at the edge, the surplus bouncing off and being drawn into the waste-disposal system. An average spraying pressure is about 275 kN/m^2. A suitable colour can be mixed with dye and industrial spirit, but in the United Kingdom a colour mixing service is available.

Most work is coloured on the head only and to achieve this without spreading the colour on to the fore-edge the work is arranged so that the colour jet strikes the head at the spine corner first. It is mainly works of reference that are coloured on all three edges but this affords little protection for the leaves.

Books that show marked thumbing on the fore-edges are often sprinkled to both decorate the book and to camouflage these marks. Work that has to be hand sprinkled is showered by the spray from a large brush dipped into red ochre liquid. An almost dry brush produces fine spray and a similar effect can be produced by hand-held spray pumps. In the spray booth sprinkling is produced by reducing the air pressure to around 50 kN/m.^{2}

Marbling is today confined to account-book work but has in the past been used very effectively on fine letterpress work. Trough marbling may be used to decorate both book edges and sheets of paper for use as endpapers and for sides.

For this process a gelatinous size is extracted by boiling Irish seaweed (carragheen moss) in water. The size is placed in a zinc tray of suitable dimensions and the surface skimmed off with paper. The prepared colours resemble poster colour in composition and to a small quantity is added a few drops of ox gall. When dropped on to the surface of the size a single spot of properly balanced colours expands to about 100 mm and different colours are laid down in spots or strips to make a pattern. A stylus drawn the short way through the colour will pull them into narrow bars and the pattern is completed by passing a steel toothed comb through the colours lengthways.

The book edge that is to be marbled is clamped, washed over with alum solution and dipped into the colour. White bubble marks, caused by trapped air, are prevented by carefully angling the descent of the edge into the size. When lifted, a certain amount of size covers the pattern, and this is removed by gentle hosing with water and the edge is then allowed to dry.

Various patterns can be produced; Spanish, Gloster, Stormant and Non-pariel are popular as well as the Dutch described above. Sheets of paper are marbled in much the same way, a modern exponent of marbling as an art form being Douglas Cockerell who devised some wonderful patterns.

Marbling is a one-time process, a new pattern being required after each sheet or group of books has been dipped. This makes it a most costly process in time and an alternative transfer method is used for commercial book edge marbling. This is effected by damping the book edge, placing the transfer against it and then damping the back of the transfer paper. In a few moments the backing paper can be peeled off leaving the pattern on the book edge. This is a very poor imitation of trough marbling and the difference in quality is marked.

Although marbling is carried out mainly for edge decoration, the fraudulent removal of a leaf from an account book with marbled edges leaves evidence, and this fact may play some part in the continuing popularity of the process.

Colouring, sprinkling and marbling provide no protection for the book and in certain circumstances this may be corrected by burnishing on top of the colour. When thoroughly dry a layer of clear wax is gently applied to the book edge. This may then be worked into the paper with the stone burnisher and a bright hard edge will result. A softer result may be had by brushing the edge with a short bristle brush; light colours should not be waxed as this tends to dull the colour. Burnishing helps to keep out dust and injurious acids from the atmosphere.

Shape of books

The cheapest hardback books, paperbacks and magazines are left square backed and during their normal lifetime these volumes nearly always take up the well-known convex fore-edge/concave spine shape. The fore-edge of squareback casebooks, with relatively light linings, on the spine, soon project beyond the case and at a very early stage in the life of the binding excessive strain is placed on the hinge which ultimately collapses (fig. 6.6). This is especially true of thick and heavy books and this shape should be avoided unless the economics of the job allow some extra linings to be added to the spine.

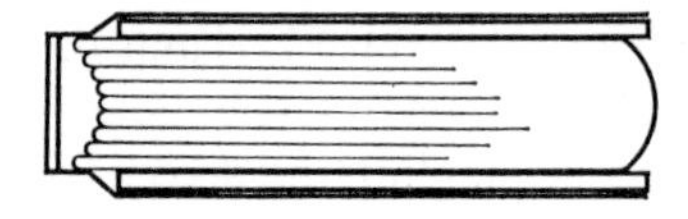

6.6 *Square back book after being in use for some time.*

Most permanently bound books are rounded and in the case of letterpress books, backed also. Rounding utilises the natural tendency of a sewn volume to drop into a shape other than square and the amount of round that can be induced into the book is directly proportional to the amount of swell. A book that has considerable swell will round excessively and one with no swell at all, *eg* side sewn and adhesive bound work, will not take up any round.

A certain fashion and some expediency is observed here. American binders appear to prefer case books of flat arc and books coming from automatic production lines also tend to be relatively flat. The hand binder will hammer the book into the required shape aiming for an arc that initially may seem excessive, but there is some loss during the later binding stages (fig. 6.7).

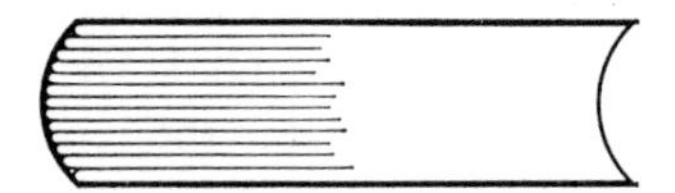

6.7 *Book with suitable round.*

Small rounding machines have a reciprocating beam to reproduce the hammer action and by forcing the spine against the action of the beam the book is made to take up the required shape.

If a sewn and rounded book is placed between two boards and pressed (fig. 6.8) the spine naturally tends to move against the force of the pressure. The backing operation accentuates this movement, fanning the sections to form a housing for the book's boards, consolidating the round and unifying the spine. The book is placed in a wooden or metal press that has bevel edged jaws; the correct amount of paper is left projecting and the spine is then beaten with the hammer into the correct shape (fig. 6.9).

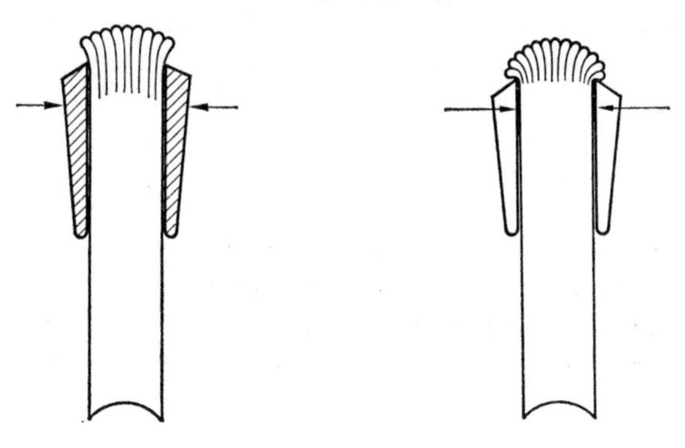

6.8 *Rounded book ready for backing.* 6.9 *Hand-backed book.*

At the end of the last century John Starr of New York devised a simple machine for backing; this consists of steel jaws and a roller to replace the hammer. With the book in position the arm carrying the roller is brought to

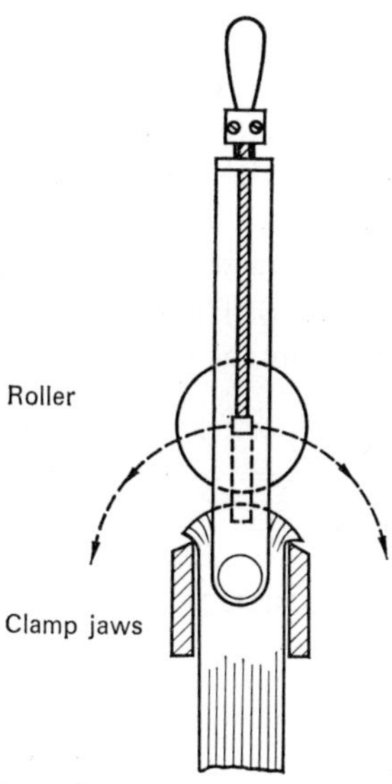

6.10 *Roller backing of thin book.*

a vertical position; by treadle movement the roller is brought into contact with the back of the book; the roller is then moved across the book in an arc to complete the sequence (fig. 6.10). Different arcs are obtained by moving the roller vertically in this arm, the highest position giving the biggest arc for the thickest book. In this instance the jaws of the machine are raised by a separate wedge system to keep the book spine in contact with the roller. These principles are maintained in modern hand-operated backing machines which have spring systems to make the task easier. In-line machines use mechanical or hydraulic methods to round the backs of books at speeds up to thirty-five books per minute.

The size of the shoulder given to a particular book bears a direct relationship to the thickness of the board and the style of the binding to be adopted. For case work the shoulder made on the book is 1½ or twice the thickness of the board to be used. This allows for free opening of the completed volume (fig. 6.11).

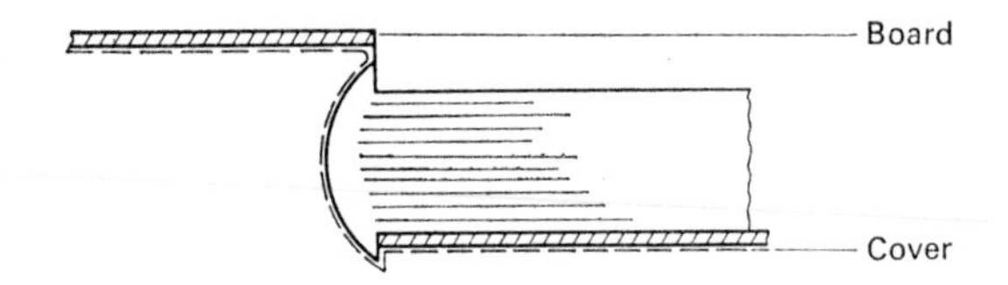

6.11 *English joint on case book.*

Using this system, very large books, demanding extra thick boards would need extra large and clumsy shoulders. As these books usually require a thicker covering material, an alternative method is adopted for the shoulder-board relationship. This introduces a groove between the shoulder and the board allowing the covering material to flex without the necessity of the shoulder being higher than the board (fig. 6.12). The shoulder for french groove books should be the same as or slightly more than the board caliper.

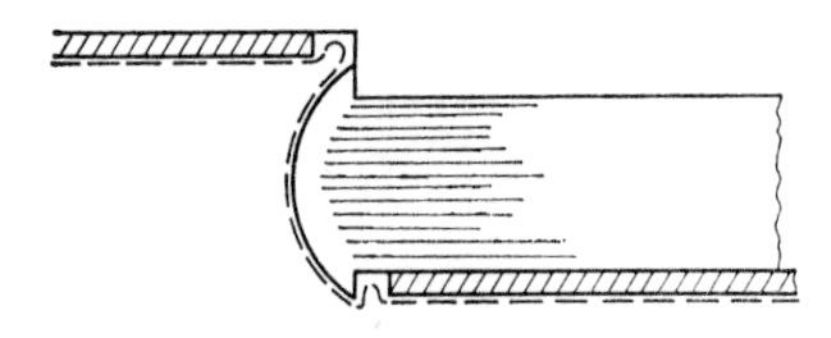

6.12 *Book with French groove.*

The backing standard described is applicable to books that are pasted down shut, *ie* the board taken to the endpaper; extra and half-extra work is pasted down open, *ie* the endpaper taken to the board, and although the boards are close up to the shoulder in these styles the board/shoulder height is commensurate. Stationery bindings are never backed as this would prevent the flat opening essential in this type of work.

Linings

To enhance the life of the volume it is necessary to support the spine folds in some way and permanently bound books have linings glued on.

If the book-covering material is adhering directly to the spine, so that it takes up the shape of the book when it is opened, the book is said to be 'fast' or 'tight' backed and often the covering material is itself sufficient support and no further linings are required (fig. 6.13). Wrappered, quarter bound cut flush, and 'yapp' styles are examples of this.

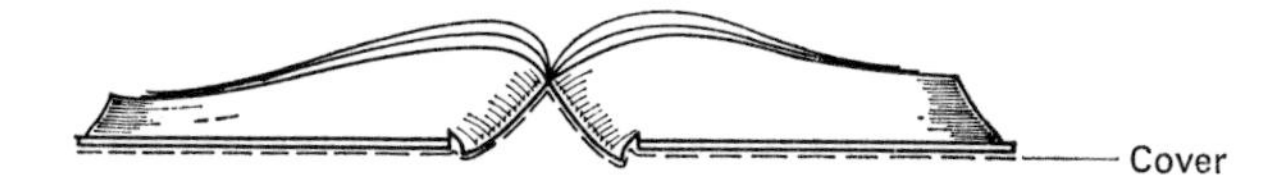

6.13 *Book with fast back.*

Papers, fabrics and leathers are used for lining, the latter only for best work and particularly good-quality ledgers. Traditional sewn and cased books have a first lining of mull and a second lining of paper each being attached by its own layer of adhesive before being rubbed firmly into place (fig. 6.14). Books that are lined before rounding and backing must be lined with a material that will expand to accommodate the extra width of the spine after backing. Expanding calico (stretch cloth) and latex impregnated papers are available for this purpose.

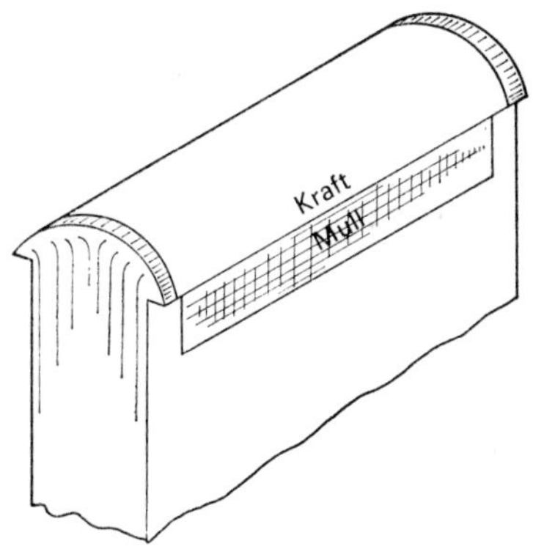

6.14 *First and second lining of a case book.*

When constructionally convenient the material used for reinforcing the spine may be extended on to the sides and in this way the hinging joint of the end-paper is strengthened. Examples of this are the mull on casebooks and the leather linings on some ledgers. The alternate layers of glue and lining material help to prevent the book losing its shape and returning to the flat condition, with the attendant faults, mentioned in the previous section. Thick books printed on heavy stock are particularly prone to this condition and many layers of linings should be applied to try to prevent it.

6.15 *Book with hollow back.*

Fast or tight back books using a stiff cover material do not open very freely. To improve the opening the linings may be arranged so that the cover can move away from the book spine as the book opens. In this way the pages are allowed to rise freely into a flat opening position; this is termed a 'hollow back' (fig. 6.15). Examples of this are case styles and account books; although the latter incorporate a specially made 'spring back' that supports the book in a flat condition whilst hand written entries are made.

Attaching boards

Selection of a suitable caliper of boards for a given book is an important item in its construction. Board that it too thin will inadequately protect and stiffen the book whilst thick boards give the volume a clumsy appearance and if the board is also heavy, place an extra strain at the hinging point. Strawboard and chipboard are used for the bulk of commercial work while the more stable and dense millboard is preferred for better quality binding.

Three methods of attaching boards are used. The cheapest books have

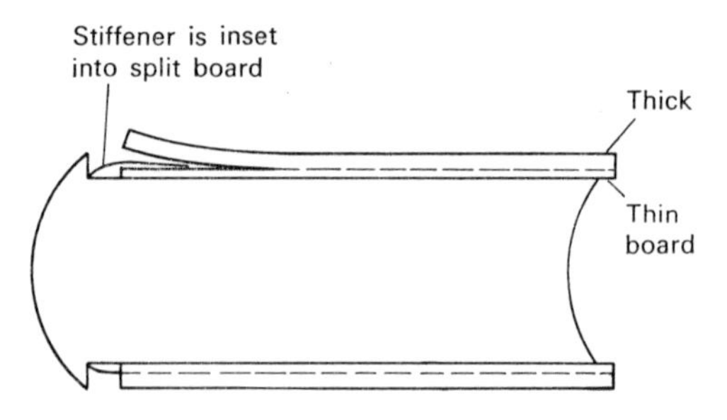

6.16 *Split board method of attachment.*

single boards glued to the waste leaf or endpaper and the strength of attachment rests upon successful adhesion. Casings and flush work are examples of this.

For middle-quality work, requiring greater strength, the split board technique is used. This is usually a thin and a thick board glued together so that tapes and linings may be sandwiched between the layers (fig. 6.16). This method is used on certain ledger binding and library rebinding work. In the latter case an imitation split is sometimes made by splitting a single board with a rotating knife.

The third method of attachment requires that the book has been sewn on hempen cords. These cords are frayed out, pasted and laced into holes punched into the book board. These are subsequently hammered and pressed so that the fibres of the hemp become part of the board and give a strong and flexible attachment (fig. 6.17).

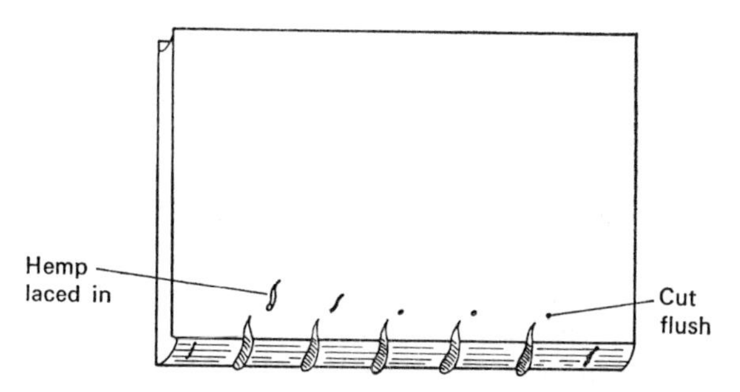

6.17 *Laced-on boards.*

The overhang of the boards or 'squares' bear a relationship to the book area. An A5 volume looks well with 3mm squares whilst A4 may need 5mm overhang to look right. Aesthetically the squares look better balanced if the fore-edge is slightly larger than the head and tail square.

Covering

The cover material is the first part of the book to be seen by the user and it should make a good impact. Leathers, fabrics, paper and plastics are all used in various ways. Three main styles are in use 1, full or whole bound, when the book is covered in only one material which extends over the whole book (fig. 6.18); 2, quarter bound, when, for aesthetic and economy reasons, the spine is covered with one material and the sides with another (fig. 6.19); and 3, half bound which has the back and corners of one material and the sides of another (fig. 6.20). It is not usually economic to use styles 2 and 3 unless the saving

6.18 *Full bound work.* 6.19 *Quarter bound work.* 6.20 *Half bound work.*

in cost of material is sufficient to warrant the extra cost of labour in cutting, handling and affixing several pieces instead of the single piece used in 1.

Work that has the board and cover trimmed to the same size as the text is said to be 'trimmed flush'. Most other work has the board edges covered by turning the material in so that it adheres to the inside of the board.

All but the thinnest leathers used for covering require to be thinned or 'pared' at the functioning points or where two layers coincide and may give the book a clumsy finish. A sharp knife is used to reduce the thickness along the edge and if extensive paring is needed the leather may be clamped and spokeshaved. A small machine with rotating cylindrical knife can be used when larger quantities are involved. Bookbinding leathers can usually be moulded and this important feature is utilised to produce the 'bands' on the spine of an extra letterpress volume and when a 'headcap' is made to strengthen the top and bottom of the spine.

Other covering materials are supplied in rolls and when cut to size the warp of fabrics and the machine direction of paper material should, ideally, lie parallel to the spine. Choice of cover is directly related to use and price of the volume; a good quality leathercloth may be thirty times more expensive than paper felt but will take hard wear for a considerable time and will have much better tensile strength and resistance to abrasion.

The adhesive chosen for covering depends upon the nature of the cover material and the complexity of the covering process. A simple task in cloth requires a tacky fast-drying glue, whilst a full bound extra leather volume taking twenty minutes to cover will need a slower drying adhesive, probably a starch paste.

To make a neat finish the corner of a cover is trimmed off and, after one edge is turned over, the surplus material on the corner is 'nicked in' to form a neat tuck before the other edge is turned in to complete the task. A stronger but more clumsy corner is made by leaving the cover intact and turning the unwanted cloth on to the board and making a 'boxpleat'.

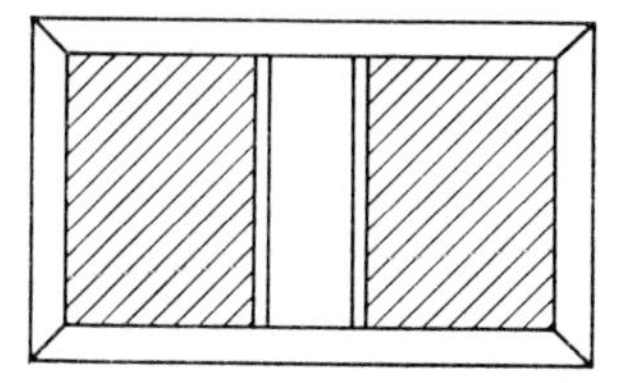

6.21 *Finished case.*

A cover made independently of the book is known as a 'case' (fig. 6.21) and is assembled from two boards, cloth and hollow paper. A single case is made on the book, but small quantities may be produced using a gauge to correctly space the boards. Assembly jigs are sometimes used for quantity work when the shape or size of the finished case takes it outside the normal range of case-making machinery. Cases with thin boards and only the turn-in of the cover glued to the board are described as 'limp' and are used for diaries, religious works and sets of classical volumes.

Pasting down

The last operation in forwarding is to attach the pastedown sheet of the endpaper to the inside of the cover. This effectively hides most of the constructional characteristics of the volume and provides a clean tidy opening. In case work a large part of the hinge strength is derived from this operation.

Usually the endpaper is coated with adhesive and the cover is lowered gently into position. Excessively wet adhesives will extend the paper and cause the sheet to project forward of the book at the fore-edge. Very dry adhesives may act before the book can be positioned in the press and for these reasons the hand bookbinder will often use a mixture of animal glue and paste for this operation. Care is taken to ensure that the adhesive does not ooze down into the book edges and that sufficient adhesive is placed in the hinge of the book to ensure good adhesion and strength.

Cases, previously made and blocked, still have a flat spine and require shaping before casing in. Care is taken when marrying up the two units to ensure that squares are properly adjusted. Extra letterpress bindings, in leather, are pasted down open, *ie* the pasted endpaper is taken to the open board and allowed to dry in the open position.

Specifications for some popular binding styles

Name	*Endpapers*	*Method of securing the spine*	*Treatment of edges*
Flushwork	Tipped, hooked or stiff leaved	Adhesive, sewn or wirestitched	Sometimes sprinkled
Common account book	Cloth-lined joints	Machine tape or web sewn	Marbled or sprinkled
Half extra account book	Cloth-lined joints	Hand or machine web sewn	Marbled or sprinkled
Case	Tipped-on 4-page	Sewn or adhesive	Coloured, sprinkled or gilt
Library rebind	Cloth-lined joints	Sewn on tapes	Sprinkled
Yapp bible	Tipped-on 4-page	Machine french sewn	Gilt round corners
Pocket diary	Tipped-on 4-page	Machine french sewn	Sprinkled, stippled or gilt
Half extra letterpress	Made endpaper	Recessed cord sewing by hand	Gilt or coloured
Extra letterpress	Made with optional cloth or leather hinges	Recessed cord sewing by hand	Gilt or coloured

Spine shape	*Linings*	*Board attachment*	*Cover material*	*Decoration*
Square or rounded only	Usually none	Single board glued on	Textiles or papers, flush or turned in	Usually none
Rounded only	Fabric with spring back	Single board glued on	Leather, buckram or leathercloth	Blocked title
Rounded only	Leather; spring back drawn on	Split board	Leather, half bound	Hand titled
Rounded and backed	Mull and kraft	Single boards in case construction	Paper, textiles or leather	Blocked or pre-printed
Rounded and backed	Mull and kraft	Split board	Quarter leather, leathercloth or waterproof cloth	Type or slug
Rounded with pinhead joints	Tight back	Thin flexible material, case construction	Full leather	Blocked
Square	Calico lined	Thin flexible material, case construction	Thin leather or leathercloth	Blocked
Rounded and backed	Hollow lined	Laced on cords	Quarter or half leather	Hand finished in gold
Rounded and backed	Hollow or tight back	Laced on cords	Full leather	Individual design tooled in gold

Pressing

All books require pressing after the endpapers are in position to ensure that endpapers and boards are in proper contact while the moisture dries out. Books usually improve considerably from a good pressing when the construction is consolidated. A minimum time of two hours is considered essential although most binders try to arrange a longer period, perhaps overnight, in the press.

Pressing boards are made from plywood or other suitable material, some having a special metal edge to press the cover material into groove work (fig. 6.22). Traditional presses are of the metal screw variety but electric, hydraulic and pneumatic presses are available.

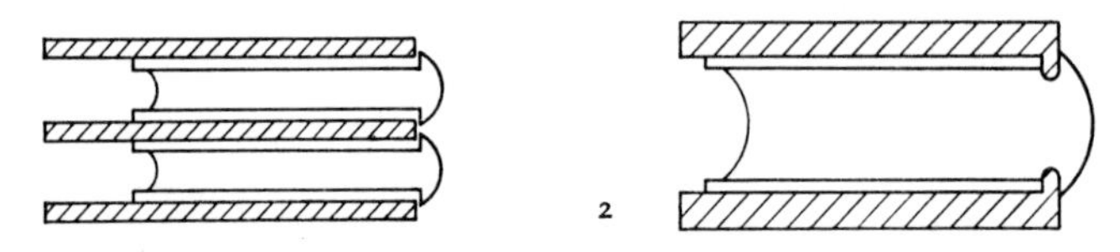

6.22.1 *Books between pressing boards (English joints).*
6.22.2 *Book between metal-edged boards for grooved work.*

Inspection

After pressing a visual inspection of the work is usually carried out to ensure that no obvious faults are apparent and that workmanship and materials are up to specified standards. Best work is often 'opened up', *ie* the cover and some of the pages are turned over to release the book and start it functioning correctly.

7. Cover lettering and decoration

A book is never complete until its title appears on the cover and frequently the plain cover of the book is also used as a medium for a design that is purely decorative or may attempt to impart the feeling of the subject matter contained in the contents. The history of bookbinding tends to some extent to be the history of bookbinding design and many fascinating volumes have been written on the subject. Book designs have usually followed the design fashion of the day with type design, architecture and furnishings all having a strong influence. Customers have always requested bindings in past styles so that the repertoire of the skilled gold finisher has to be very wide indeed.

Lettering and decoration by hand

The traditional hand operations continue to be used with little change over the years. While single books require lettering these methods will continue to be used, the problem and its solution remaining unaltered. To be effective the image must be impressed below the surface of the cover so that normal handling will not wear away the very thin metal; also the grain of the material has to be 'bottomed' to give the impression a solid appearance.

Gold leaf is the traditional medium used for the purpose of lettering and decoration because it is fairly easily worked into thin sheets and does not tarnish in use; books several hundred years old in libraries and museums are evidence of this. Also available, but not often used are silver leaf, which tarnishes, and platinum leaf, which is expensive.

Because of the extensive preparation work needed with gold and other leaf metals its use has declined in favour of ribbon gold and other metallic foils. Only the very best bindings, single books and those books bound for archive libraries, when longevity is absolutely essential, are now tooled with gold leaf. Ribbon foil appeared between the two great wars and its use greatly reduced the time factor in tooling; these are discussed in the section on blocking.

Tools used in hand work

A range of hand-held engraved brass designs are used in combination to create the desired effect. Single letters of various sizes, logotypes, decorative pieces, lines and engraved rollers are heated and impressed into the surface in the required pattern (figs 7.1/2/3/4).

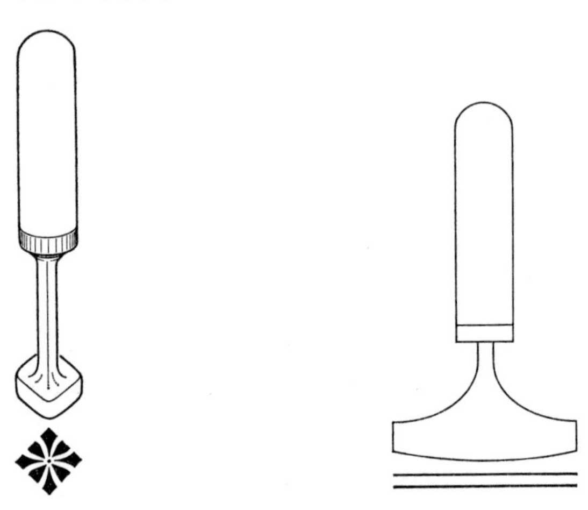

7.1 *Hand-held centre tool.*

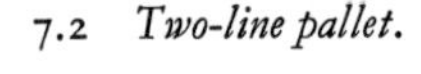

7.2 *Two-line pallet.*

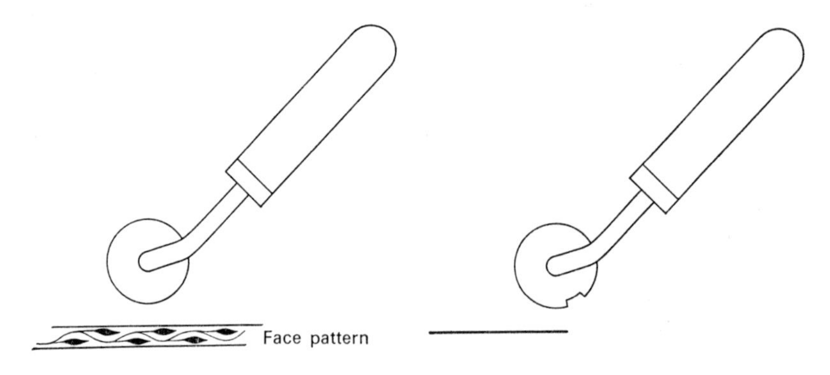

7.3 *Engraved brass roll.*

7.4 *Fillet.*

An alternative method of titling is to assemble type into a hand-held type-holder. Tremendous pressure is needed to impress a long line below the surface of the cover and the length that can be worked into a flat surface is limited; impression into a curved book spine is easier. Brass type is available in many designs and sizes and where appropriate founders and 'Monotype' type, and Linotype and Intertype slugs may be used. It should be observed that brass type resembles printers' titling founts and when printing type is used in the typeholder the head of the letters must be very close to the top of the body for sighting purposes.

The only other limitation in the design of the face of type and handle letters for finishing and blocking is the thickness of serifs and strokes. If very fine, these are apt to be lost in the heavy grains used for book cover materials.

Method

Tooling in leaf gold requires that the surface to be worked is coated with a thin solution of egg albumen which is allowed to dry. The leaf gold is laid on to the surface which has been previously lightly greased.

Two basic methods of marking up for hand tooling are used. When a complex design is to be tooled it is constructed on thin bank paper and when the right effect is achieved this template is laid over the cover and the design tooled through the paper and into the cover material. This indirect method produces excellent results but requires a double run to complete the tasks; the first in 'blind' impression and the second in gold.

The direct method requires the cover to be prepared and the gold laid on. Then by using a sharp pointed marker, scratch lines are drawn into the gold to act as guide lines to the finisher. If, for instance, a title is being tooled, equal spaces will be allotted to each letter in the marking guide and the finisher will adjust the spacing visually as the letters are being tooled in. As corrections are exceptionally difficult to make, high degrees of skill are exercised to achieve alignment, verticality and correct spacing of every individual letter in the line.

When the heated tools are impressed into the gold the heat is transmitted to the albumen (glaire) which becomes tacky and secures the gold leaf to the surface in the design of the tool. To complete the task, surplus gold is rubbed or washed off with spirit.

A similar technique is used when ribbon gold is used for tooling but all the guide lines can be scratched into the ribbon before it is taped into position on the book. In this case no previous preparation of the cover material is normally necessary.

Blocking

This process, developed in London by Archibald Leighton, in 1825, was designed to speed up the production and improve the quality of publishers' cases. Today its application far exceeds the narrow confines of bookbinding and is variously called 'hot press stamping' and 'marking' in different industries.

Hand-operated blocking machines are usually designed to function so that the surface being blocked rises vertically on to impression (fig. 7.5.1). The

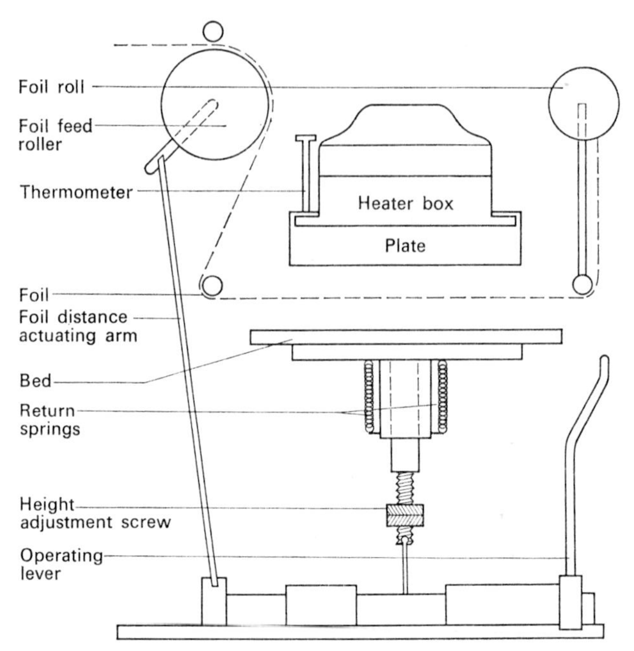

7.5.1 *Hand-operated blocking machine (front section).*

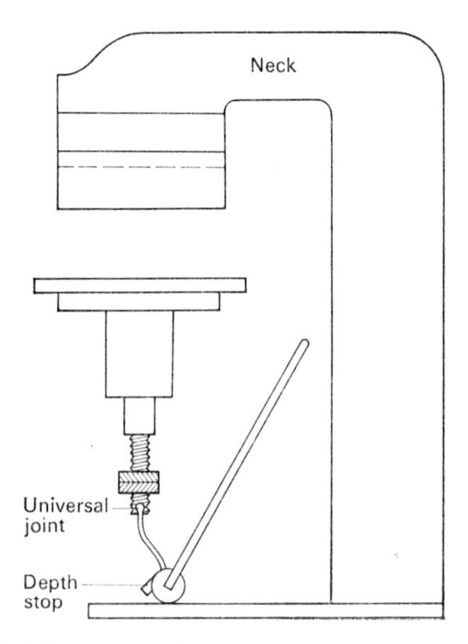

7.5.2 *Hand-operated blocking press (side section).*

blocking plate or type is mounted horizontally, face downwards, to a heated box and in a position as close to the centre of the machine as possible. The slide carrying the book cover is supplied with suitable gauges and when pushed forward the cover is correctly positioned beneath the block or type. Ribbon foil may be cut from the roll and laid on but when quantities are being processed a foil feeding attachment may be used; this is adjusted to move the foil forward in lengths suitable to the job in hand. When the actuating lever is moved the cover is made to rise up into contact with the heated plate, where heat transference and impression occurs. A depth limit stop is usually incorporated so that correct depth of impression is maintained.

Blocking foil

This important material is composed of four layers; the carrier, release wax, coating and sizing.

The last three are carried by a roll of fine paper or film and it is important that this material be both thin and strong. Thin foil helps to maintain definition while the thicker foil tends to thicken up the strokes of the image. Tensile strength is important when the foil is being run at high speeds in an automatic machine and when breaks might be expensive. Glassine is a thin rather brittle paper and is used extensively for the cheaper range of foils. Viscose is a cellulose-based film with greater tensile strength which tends to curl rapidly when heat from the press dries off the moisture content. Polyester film is the most recent and successful of the carriers used; it is extremely thin and has great tensile strength. Its use as a carrier is being steadily developed and most of the coatings are now available on this material. A typical modern blocking foil based on polyester may caliper as little as 0·0125mm.

The release wax is interposed between the carrier and the coating and its function is to melt and release the coating from the carrier at about or a little below the temperature needed to set off the size adhesive; this should allow a clean separation so that little or no cleaning off is necessary after impression. Considerable advances in the quality of release waxes have been made in recent years.

Metallic coatings are atomised on to the prepared carrier film in vacuum chambers by a continuous vaporisation process. Gold is used in carat ratings from 21 down and by using different alloys slight variations of colour are obtained. This may be important because in certain cover materials some gold colours appear brassy or dull while another may give a satisfactory 'gold' appearance. This coating is used for all hand finishing and the best of blocked work that must be guaranteed 'pure gold' or that which needs to stand the test of time. It will not tarnish and correctly applied will look rich and satisfying.

The introduction of aluminium foils and the tinted and dyed varieties has added tremendously to the range and scope of the blocking process. A wide range of gold and metallic colours are now available and have been tested for light fastness representing many years of normal exposure and have proved to be very durable. It is extremely difficult to separate, visually, real gold from tinted aluminium of the same colour on a completed job and today a high proportion of the blocked book titles and decoration is completed in this material. It is also widely used for the decoration of boxes, cartons, letter-heads, greetings cards, showcards, plastic packaging units, pencils and ball-point pens, shoe inserts and hat bands, etc.

The cheapest and least satisfactory of the metallic coatings is the deposition of aluminium and bronze powders to the carrier film. These give a flatter, less brilliant appearance than gold or aluminium and are quite easily recognised visually. Bronze tends to react with the cover material forming a green or black oxide and this may happen within weeks of the book coming into use, and for this reason these foils are now seldom used for book blocking.

Pigmented foils are available in a wide range of colours, matt or gloss finish. These tend to be thickly coated and image strokes thicken up in blocking.

Sizing

The last layer appears on the underside of the ribbon and shows as a white, tan or yellow deposit. The size acts as the adhesive and is heat reactivated. Blocking foils are sized either for blocking on fibrous materials, *eg* paper, leather, cloth and board; or for impervious plastic surfaces such as polyvinyl chloride, polyolefin, etc. It is nearly always desirable to carry out a test run and prove the foil whenever a new cover material is encountered.

Purchase

Foils are made in rolls up to 900 mm wide and 610 m long (except gold, which is 275 m). These are supplied cut in widths suitable to the job or may be cut in the bindery on a small lathe-type machine. Gold ribbon is about four times the price of aluminium taking into account the different lengths of the reels.

Blocking plates

The most widely used blocking plate is the engraved brass. This is produced by photographing the prepared art work and printing down on a suitably prepared brass plate of 6·35 mm caliper. After a brief chemical etch the plate is deeply routed and hand engraved. The depth of cut is important as shallow surfaces will make contact when on impression and counters will fill in.

The less expensive engraved copper plates are useful when a shorter run of good quality is to be worked. Binders' electrotypes have a thicker shell than the printing variety, but the soft backing metal renders the plate liable to rapid wear unless treated gently by the press operator. Engraved steel, plastics of various types and zincos have their uses under certain conditions. Sometimes blocks are plated with nickel or chrome to extend their life.

Principles of blocking

Three variable factors control the quality of blocking: heat, pressure and dwell. The heat used will vary with the materials being employed and may also be affected by the build up of heat caused by the block being in contact with the surface for an extended period, *ie* dwell time. An average thermostat setting of 100°C may be considered a good starting point. The temperature at the face of the block varies from that registered on the thermometer dial and is subject to loss due to draughts and high operating speeds. Empirical methods are used to arrive at the correct thermostat setting for a given job which, in any case, may vary during the day or from day to day. Simple hand-operated machines are often deficient of thermostats and temperature control is effected manually.

Pressure on machines is applied through an operating lever and mechanism which gives a fixed rise and fall, variations in plate height being compensated by an adjustable screw linkage or pressure wedge beneath the platen. By correctly setting the pressure, the plate is made to bottom the grain of the material; light grains may require a medium pressure and short dwell to achieve the desired result; heavy grains require the softening effect of longer dwell allied to correct pressure to obtain the same result. (fig. 7.5)

Dwell is simply the time that the plate stays on impression. On some automatics this is adjustable, but on hand-operated machines it is largely a matter of judgement on the part of the operator. Excessive dwell results in heat build up, filled counters and blurred outline. Insufficient dwell gives no time for the size to reactivate and/or cloth to soften and this gives an impression that lacks solidity or allows the grain to show through the coating. When very heavy grains are to be blocked it is sometimes desirable to 'blind' run the job first. This is a run of the job without the foil in position and commences the process of grain flattening. In some instances the attractive texture variations obtained by blind blocking are made a feature in the decoration of the cover.

Blocking machines vary widely in size, type and principle. Hand-operated presses are relatively cheap but slow in production. Hand-fed power machines include mechanical and pneumatically operated types; the latter are often chosen for blocking materials that have inconsistent caliper and texture throughout the stack, *eg* leathers. The mechanism on these can be set to continue to apply pressure until a certain standard is reached; in this way the machine 'feels' its way into the surface tending to treat each new surface individually.

The more sophisticated automatic platen, vertical and rotary machines are discussed in the chapter dealing with publishers' binding.

8. Box and carton manufacture

Three main areas of production can be identified: 1, corrugated and fibreboard containers; 2, rigid boxes; and 3, cartons. The use of various plastic materials in packaging is growing fast and supplementary units for paperboard containers are used, *eg* polystyrene mouldings, vacuum-formed platforms, flexible wrappings and laminates, etc.

Corrugated and fibreboard containers

These offer maximum protection for packing glass and other breakable goods, for the transport of large objects, and are widely used as a final pack for conveying quantities of small goods to stores and shops.

The principal method of production is to feed the container blanks into a machine that combines printing and slotting. Rubber stereo plates are the usual printing medium used, in conjunction with aniline inks, and the sheet of printed material then passes directly into the slotter unit.

Here a series of adjustable rotary punches and counterparts set across a shaft cut away the material to form slots that give the shape to the flaps of the finished container. At the same time creasing discs put the lateral creases in and form the sides (fig. 8.1). On these machines, speeds of up to 15 000

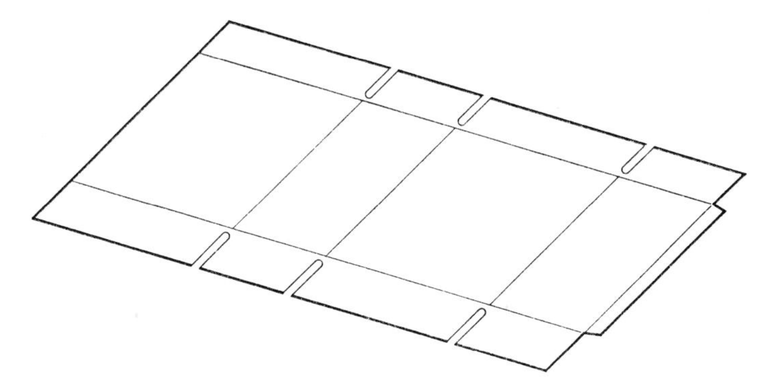

8.1.1 *The container blank is cut, creased and slotted.*

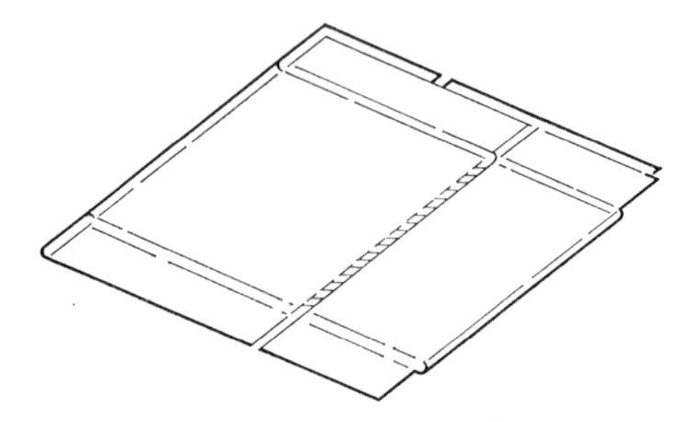

8.1.2 *The stitched blank is left flat for delivery to the customer.*

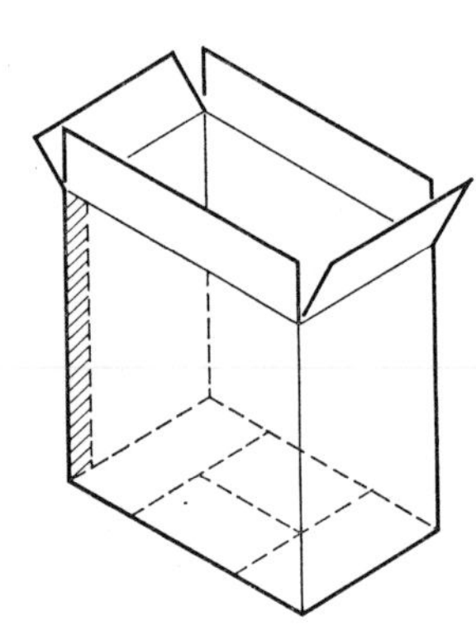

8.1.3 *The container erected with base flaps glued.*

blanks per hour can be achieved depending upon the quality of the printing specified.

Alternative methods utilise conventional cutting and creasing dies in conjunction with platen presses.

Make-up in this section of the industry includes folding and partial closing of the construction by gluing, wirestitching or taping, using suitable hand-fed or automatic methods.

Rigid boxes

The market for rigid boxes has remained more or less static in recent years and while they are still in considerable demand almost all the advances in production techniques have been and still are with the ever growing carton and container industries. As the name implies the boxes are made and stay rigid and in this way offer maximum protection for smaller objects, *eg* jewellery boxes and gift presentation packs for the cosmetic, perfumery and confectionary industries. These boxes have very good counter display properties and can be made to look very attractive indeed when combined with satin, ribbon and both polystyrene and vacuum-formed platforms.

The four stages in the manufacture of rigid boxes are 1, cut and score blanks; 2, corner cut; 3, corner stay; and 4, cover with paper (fig. 8.2). Cutting and scoring the blank is usually done on a rotary cutting and scoring machine having a series of rotary drivers set along a shaft to the various measurements required. A sheet of board is fed into the rollers and as it passes through it is cut and scored into long strips. The machine is reset (if the box is not square) and the other measurements cut. For larger quantities a double rotary machine with right-angle cutting and scoring assemblies produces finished blanks in one pass through the machine.

The unwanted corners are next cut from the prepared blank. Powerful machines will process one or two corners simultaneously, 4 to 6 blanks thick. The third stage requires that the blank be erected to its three dimensional shape and held there by gummed tape at the corners. The tape may be glued, water reactivated or heat reactivated using a thermo-plastic adhesive and are processed one, two or four corners at a time.

Finally the box is covered with paper either by hand or machine. A wide variety of stocks are used as cover materials, including printed arts, glazed printings, self-coloured papers, fancy box coverings and leatherettes. These are cut to size and suitably shaped to provide turn-ins and mitre corners; a film of glue is applied to the whole inside surface of the paper which is then correctly positioned on the erect box and drawn round. The long edge is brushed up first and then on the short edge the overlap is turned around the corner before the shorter edges are also brushed up. The turn-in is now rubbed down to the inside of the box to complete the process.

In a mechanical method of box covering a 'wrapping' machine is used, the glue being previously applied to the cover by a sheet-gluing machine. A specially prepared block is mounted on the machine and will fit exactly into the open rigid box which has the glued cover correctly positioned to its base. The movement of the machine pushes the box and cover downwards, the cover being drawn over, tucked in and ejected.

Cartons

The carton section is without doubt the largest in the packaging industry. Everyday household things such as toothpaste, soup, tea, sugar, cereals and cosmetics are purchased in a carton. This section has grown and continues to grow as penetration into the retail market becomes more complete.

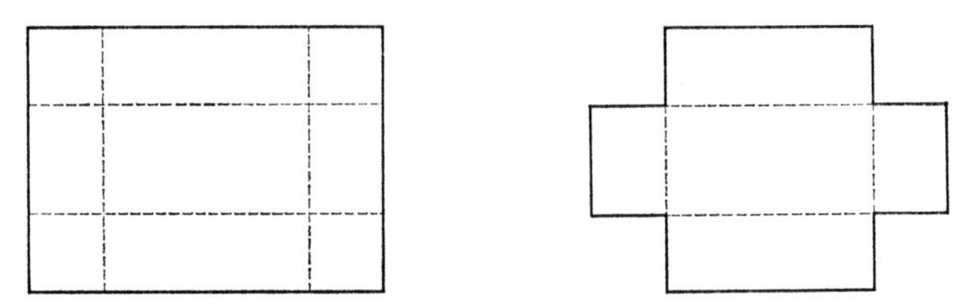

8.2.1 *The rigid box blank is cut and scored.*

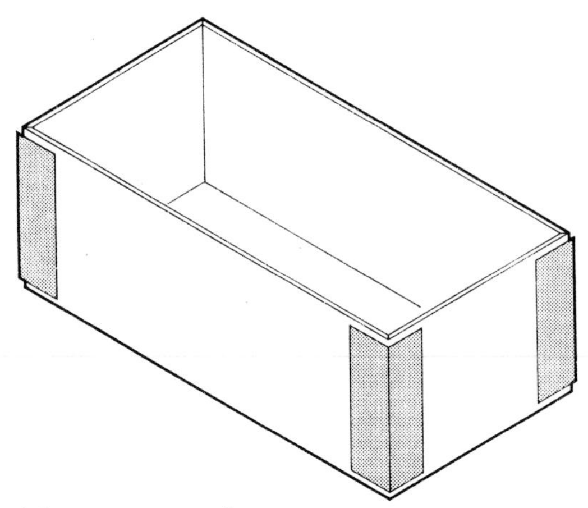

8.2.2 *The erected shape is corner stayed.*

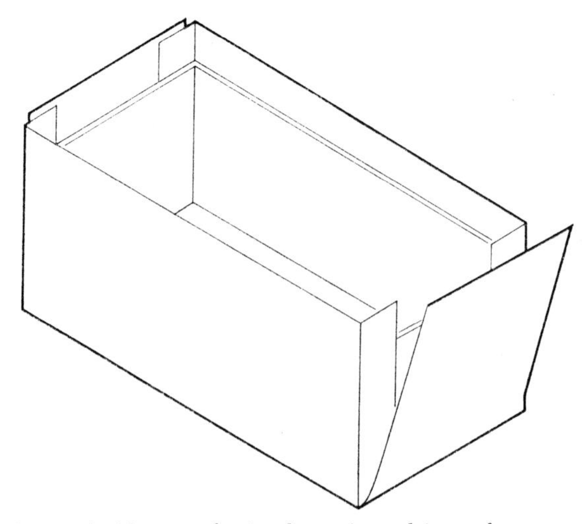

8.2.3 *The box is covered with paper, showing the opening tuck incomplete.*

A carton usually starts its life as a hand-made sample into which will fit the object that the customer wishes to pack. At this stage particular details need to be studied carefully in order to ensure that the customer will get the most suitable and economic carton. Factors include the weight of the goods to be packed; the style of the carton required or best suited to the needs of the product; economics in relation to sales; printing method to be used and the quality of the board suitable for the process.

Having studied these points and decided the style of carton required, a suitable caliper and quality of board is selected and a carton made. This will be of correct size for the measured object and make allowances for leaflets and any other items which may have to be included. Making optimum use of the grain direction the complete shape of the carton is drawn in reverse on the back of the board to be used and the lines indicating creases are actually creased by using a creasing rule and stick (fig. 8.3). The shape of the carton blank is then cut out with a cutting knife and, where necessary, pieces of cutting rule, bent to specific shapes. The blank is then checked for size and folded, glued or stitched as required. The goods are fitted in and if satisfactory the sample is submitted for approval. The sample maker will keep a register showing size, style, board used and sample for future reference.

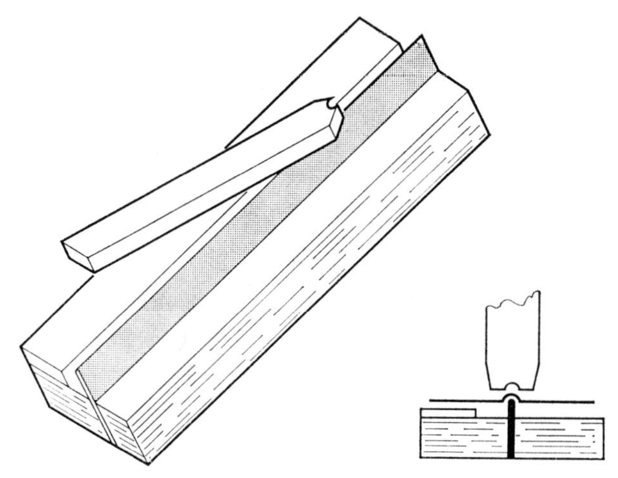

8.3 *Creasing rule and stick.*

Once the carton style and size is approved, a one off master cutting forme is prepared to the exact sizes; samples cut from this forme, together with accurate drawings are supplied to the artist to prepare the artwork. It may also be, in answer to the customer's request, that a 'pilot run' is cut from this same forme to test the carton's suitability for automatic packaging machinery (fig. 8.4). When finally approved the job goes for platemaking and printing, proofs being produced in colour on the correct material, cut to shape and a complete printed carton presented for approval. A layout sheet can now be prepared showing the number of cartons to be on the sheet correctly in position and this should agree with the original estimate. Only a step plan need be prepared showing the number of cartons in each direction of the sheet and the actual distance apart as these will be printed down by a step-and-repeat machine. From these drawings and dimensions the diemaker will make up the forme, or die, and a carbon copy taken on a sheet of astrofoil or some other stable material. This, in turn, is forwarded to the printer with which to position his plates. This ensures absolute register to the carton blank.

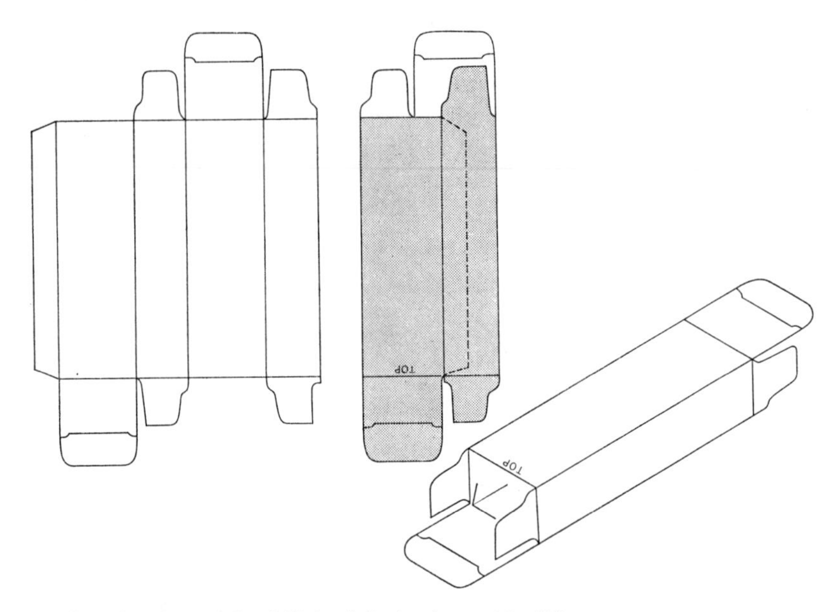

8.4 *Paperboard carton flat, folded and glued and erected for filling.*

Cutting and creasing dies generally fall into three categories: 1, metal die; 2, wooden block or multi-piece die; and 3, wooden one-piece die. The metal die, although being used less and less, due to its unsuitability on modern presses and the advancement of plywood die production, is made by composing the sizes required with metal quads and furniture in a similar way to a composed type of forme. The rules are cut to size, shaped and inserted where

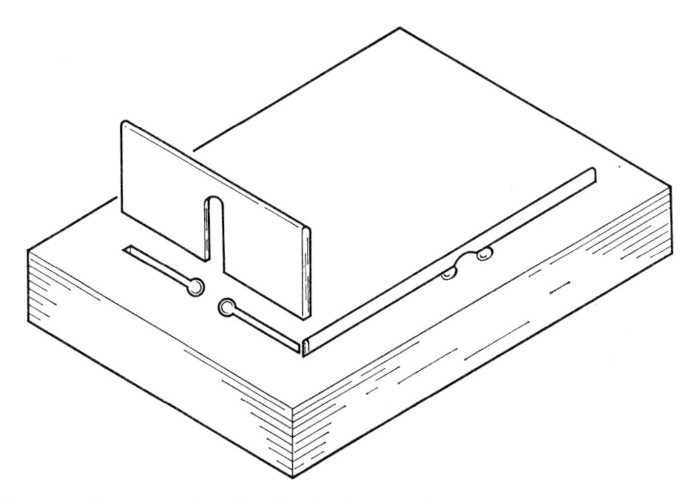

8.5 *Cutting and creasing rule is slotted to fit the sawn plywood block.*

needed. Block or multi-piece dies are similarly made, the metal furniture and quads being replaced by blocks of special caliper plywood cut by circular saw.

The wooden one-piece die is by far the most extensively used and is made in the following way. The carton shape is drawn on to plywood and this has to be repeated for the complete number of cartons. This means that if a particular job is 20 cartons on a sheet then all 20 cartons are drawn on to the surface of the wood by hand, correct to size and in position to fit the printed sheet. To prevent the plywood from breaking into a number of small pieces when the rule slots are sawn, small areas (bridges) are left; the position for these is now selected and marked. Drill holes are made at strategic points to enable the jigsaw blade to be inserted into the wood and the drawn shape is jigsawn around.

Rules are then cut to size, shaped, bridged and inserted into the wood. As bridges have been left at the side, top and bottom of each block the rules will be slotted or bridged to allow a good fit over the uncut portion of the wood block (fig. 8.5). Rules are made in many varieties and thicknesses; whilst there is a standard height for cutting rules (2·38mm) the height of the creasing rule is governed by the caliper of the paperboard to be used. The thickness of the cutting and, particularly, the creasing rule are also relative to the caliper of the paperboard from which the carton is to be made. The following is an approximate guide to rule selection but this will vary slightly depending upon machining techniques and other variable factors:

Table 1 Selection of rules for carton work

Board caliper micrometres	*Rule thickness in printer's points*
below 300	1½
325–500	2
525–750	3
775–1000	4

The die maker has at his disposal various tools and equipment to enable him to produce an accurate die. The saw is a machine combining the facilities of a circular saw, reciprocating jigsaw and rotating drill. With the aid of gauges, micrometer adjustments and quick release mechanisms the machining of the one-piece block can be effected rapidly and to a high degree of accuracy. Surplus sawdust may be drawn off by vacuum. A metal guillotine for cutting

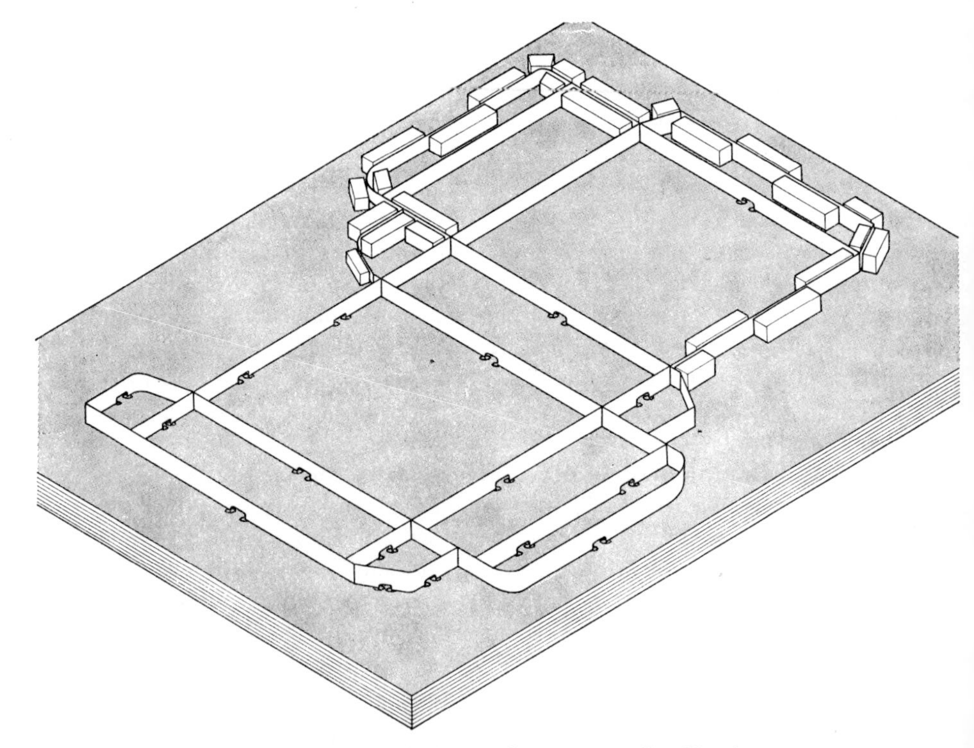

8.6 *One-piece cutting and creasing die for a single carton partially rubbered up.*

the rule to random or repeat sizes is followed by various small machines for bending, bridging and lipping as required; a grindstone is used for small adjustments, the assembly being completed on a composing surface (fig. 8.6). Tolerances of ± 0·4mm are needed to ensure that the completed cartons fit the customers packing machines.

An alternative to hand drawing is the reproduction of a well-drawn master copy by photographic means or by use of the Aldridge layout or similar machine. The positives produced are then printed down by step-and-repeat techniques on to plywood that has one surface prepared for the photographic process; this produces a fine black line on a white background. In this way faults that are associated with hand drawing are eradicated.

To check the features of the die a carton is produced from each part of the die by laying on the carton board and gently tapping around the shape and finishing off with a creasing stick. The resultant carton can be checked for squareness and ease of tuck-in.

At this stage the prepared die is 'rubbered' up, *ie* specially prepared foam rubber is placed alongside the cutting rule in steps about 6mm wide. When pressure is applied the rubber is compressed and as the pressure is released the rubber will resume its normal height and in doing so will force the board away from the rule.

Of the many types of machines being used for the cutting and creasing of the carton, three of the better known and more commonly used are described as hand or automatically fed platen, cylinder and autoplaten. The platen is used mainly for short runs and is usually hand-fed. This machine will produce approximately 800–1 000 impressions per hour on stocks up to and beyond 2 000 micrometres caliper. Showcards are also cut on the larger type platens.

The cylinder machine is usually fitted with an automatic feeder and runs

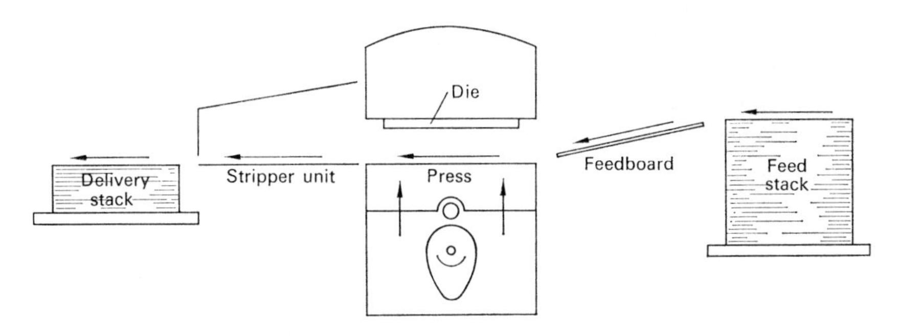

8.7 *Elements of the cutting and creasing autoplaten.*

up to speeds of 3500 impressions per hour; it is used for longer runs on medium caliper materials.

The autoplaten has the platen in a horizontal position and is fully automatic at speeds up to 6000 impressions per hour (fig. 8.7).

The preparation and makeready varies on each type of machine but fundamentally has the same object which is to apply a patched-up sheet to level the cutting impression and to ensure that all cutting knives are cutting evenly: a stencil matrix provides the female part that allows the creasing rule to actually crease the board. The makeready on a platen machine is as follows: the die, having been rubbered and checked for size and squareness is locked in a machine chase, placed in the machine and secured in position. Pressure of the cutting plate (which is fixed to the platen) is adjusted by two eccentrics fixed to the side arms and by this means pressure is applied until the highest knife is cutting. A sheet of the material to be used is placed on the cutting plate, held by a piece of gummed tape and an impression taken; under the cutting plate a sheet of kraft or similar stock is placed. When correct pressure has been reached a sheet of kraft is also fixed to the cutting plate, covered with a sheet of carbon paper and an impression taken; this will show the complete die profile. The sheet of board is then visually checked to see which knives need patching to make them cut clean; the patched sheet then replaces the sheet underneath the cutting plate in register with the die. Further pulls are made and patching effected until all knives are cutting evenly.

When the cutting is even the next stage is to produce the creases. On the cutting plate is glued a sheet of manilla or other hard paper of the same caliper as the stock being cut; this is covered with carbon paper and an impression taken. A die profile is now produced on the face of the stencil material; by cutting a line either side of a creasing rule impression and by taking away the piece between the cuts a channel is produced into which the creasing rule will force the board when the machine is on impression.

As a general guide the width of the channel should be a minimum of one thickness of the board plus the thickness of the creasing rule and a maximum of two thicknesses of board and one thickness of the creasing rule, the width of the channel chosen being one that produces the best results on the material being cut.

After the sheets have been cut and creased the cartons have to be broken or 'stripped'. In this operation the cartons are usually stripped out by hand leaving the waste area behind. On the modern autoplaten machines a certain amount of stripping can be done while the sheet is passing through the machine but this is mainly on jobs where the run is sufficient to warrant the time spent on setting the stripping unit.

The stripped cartons are then ready for the make-up operations of gluing, stitching, window patching, waxing etc. In the case of, say, a carton to hold a tube of toothpaste the side seam is glued on the carton-gluing machine. Firstly, the flap of the carton is folded over 180° and brought back to the flat plane; simultaneously the third crease is similarly treated (fig. 8.8). The glue flap receives a line of glue as it passes over a glue wheel and the carton is folded by bars and belts so that the glued flap half is stuck to the inside of the side panel. A belt ensures that the glued joint is under pressure whilst the adhesive sets; this belt travels much slower than the previous part of the machine and the cartons are delivered on top of each other and slightly staggered.

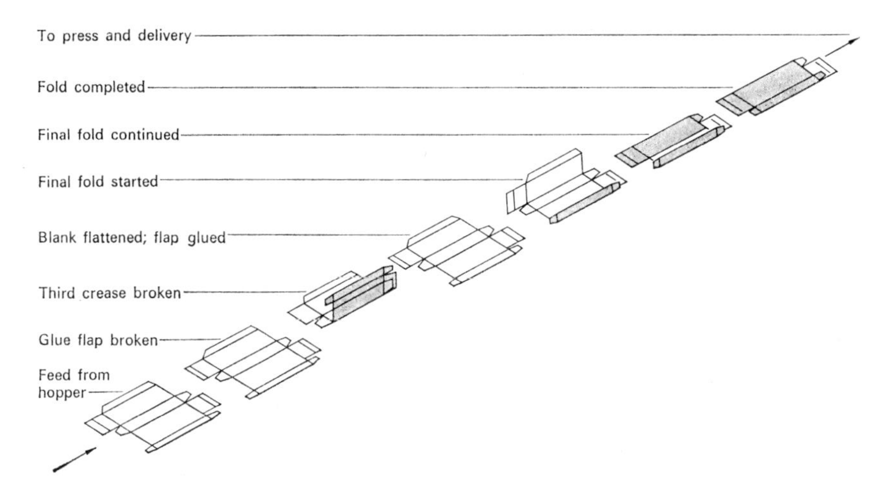

8.8 *Successive stages in the bending, folding and gluing of a typical paperboard carton.*

As the cartons emerge from the machine they are mechanically counted, gathered up by hand and either elastic or paper banded, or stacked into trays before being packed into parcels or corrugated containers for delivery to the customer. The finished carton must be carefully protected from damage; this is particularly important if the cartons are to be erected on a packing machine.

9 Packing and despatch

It is important that printed and completed work reaches the customer in acceptable condition and for this reason a satisfactory packing system must operate. The department will receive completed work from the finishing section, count the number supplied, determine the delivery instructions and pack or box the product to meet the required standards. In some instances specific instructions for delivery are written into the order or contract, but often the printer establishes his own standards within the accepted trade customs. A high standard of quality in this field should be the target as the customer is most likely to be impressed, or otherwise, by the condition of his work on arrival at his warehouse. Compact square parcels of convenient size, well taped and labelled on one end using perhaps a designed label in the printer's own house style must impress the customer and give a clear indication of the service and efficiency he can expect. Sloppy unkempt parcels that easily break, spilling the contents, are no advertisement for good service.

Packing completed work may serve a number of quite different functions. In some instances work is given a single wrap to act as a protective covering whilst it lays in a store awaiting further movement, *eg* single copies that will be 'picked' from parcels of publisher's case books for the make-up of retail orders; or holding pre-printed posters in stock ready for overprinting at a later date. In these cases the wrapping serves no other function than to keep the stock in good condition while in the store. Small or thin units are frequently banded in convenient handfuls for ease of counting and convenience of handling; the banding material should be strong and well secured with gum tape or adhesive (fig. 9.1).

Parcels that leave in the printer's own van, are posted or collected by a carrier, may require quite different treatment. This usually involves the provision of one, two or more inner wraps and perhaps even thick board strips at strategic points to give special protection to certain parts of the contents. The inner wraps are often soft bulky materials with little strength and include the corrugated type materials; some printers use waste printed sheets for this purpose although this is not specially suited to the task.

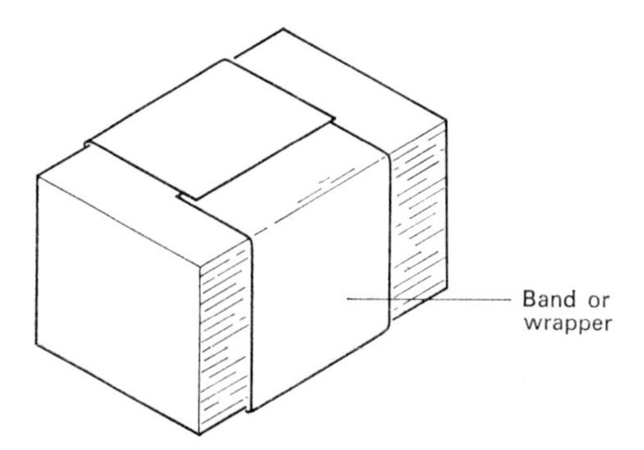

9.1 *Cut work with single-way band.*

While the printer's own delivery personnel can be expected to handle parcels with reasonable care this may not be true of either the postal or carrier services and particular care will be needed to ensure good protection of the contents of these parcels. Stringing will certainly ensure that the parcel will stay wrapped in transit but the contents may show string marks if inadequate protection is afforded; knots should be on the walls of the parcel. Weights of parcel units are determined by a number of factors; size and bulk of the contents, weight that can be comfortably lifted by the packer (often controlled by union agreements), carrier and postal regulations, etc.

The miscellaneous character of some packing operations makes for difficulty in mechanising but when large quantities of similar units are to be despatched several methods are available. 'Stringing' is the operation of tying periodicals and newspapers into bundles for bulk delivery. The bundle is assembled with the correct number of copies and standing on a waste sheet; the top waste, which already has a label affixed, is added and the bundle is pushed under the horseshoe-shaped arch of the machine. A length of string is drawn around the bundle and the ends tied under pressure; the mechanical part of the sequence takes only a few seconds and bundles may be tied in both directions.

Magazines, booklets and newspapers are often circulated by post and although hand-banding and labelling are satisfactory for smallish quantities, mechanisation is increasingly applied. Machines will flat wrap, fold and wrap or band, and some can be adjusted to handle all three (fig. 9.2). The wrapper is fed from a pile containing up to 7000 flat wrappers and the book to be wrapped is also hopper fed. If this is to be folded, guide bars create a soft fold parallel to the spine so that minimum damage is done to the ink and paper surfaces. As the book moves through the folding system pre-adjusted pressure ensures a compact unit ready for wrapping. Bands of cold glue are

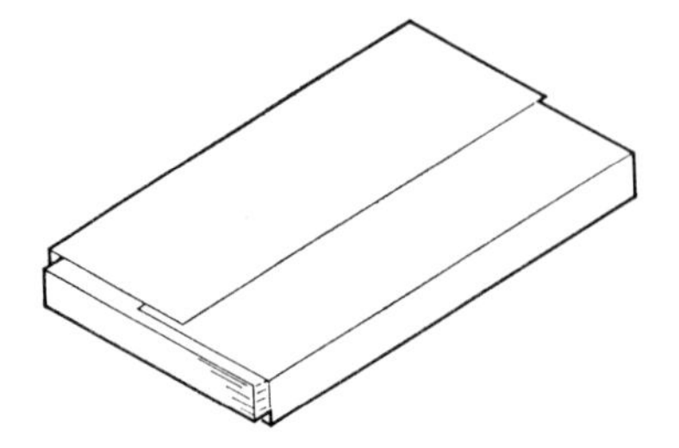

9.2.1 *Flat-wrapped book.*

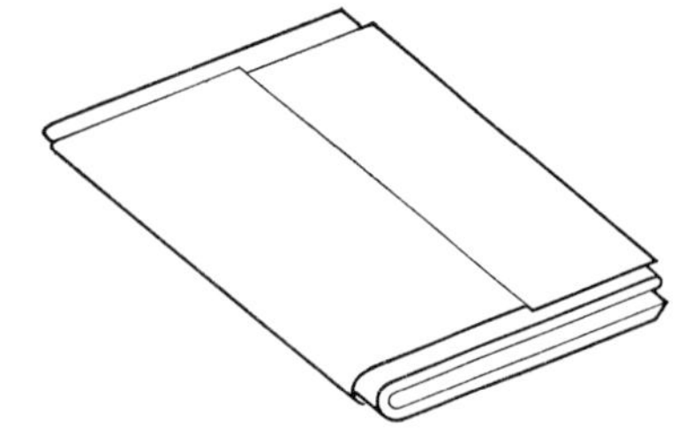

9.2.2 *Folded and wrapped magazine.*

spread on to the wrapper which is then tightly drawn around the book, folded or flat, by friction belts. Pressure rollers ensure good adhesion whilst the book passes from wrapping to the (optional) addressing head.

Addressing may be by pre-addressed wrappers, printing head or more frequently with pre-addressed labels supplied in reel form by the publisher. This type of machine provides an open-ended wrap or band on work up to 40 mm thick at running speed of up to 8 000 copies per hour.

The packing of reams, pads, books and other regular shapes in parcels with tucked ends can also be completed mechanically. The wrapping paper is reel-fed and wet adhesive ensures a perfect seal up to 1 000 cycles per hour. Other patented packs to accommodate single items use corrugated papers with adhesive sealing systems or pockets padded with shredded paper waste or foam and these give resilient packing to protect the contents at high sealing speeds.

A more recent introduction to the packaging of print is that of shrink wrapping. Three basic elements make up the system; a form seal machine, a shrink tunnel and a reel of plastic film such as PVC, polyethylene and polypropelene. The process requires the operator to pack the printed material loosely in film on the form seal; this is done by inserting the stack into the open side of the pre-folded plastic film and sealing the perimeter under a heated sealing bar. The package is then conveyed through the heated shrink tunnel where the film shrinks down to the contour of the contents producing a skin-tight fit. The film provides a tough skin that will not easily fracture, maintains the moisture level within the package and is almost completely dust proof; there is no protection from light and subjects likely to fade should be suitably screened. For goods that are to be retailed this visible style of packing is indispensable, labelling may be rendered unnecessary and in some fields economies gained.

Labels affixed to parcels of printed matter should show clearly all the necessary relevant details. A distinctive house label can be written or typed but a superior finish is obtained by the use of a small printing attachment for this purpose. This has a central hand- or power-operated cylinder that ha s number of transverse channels into which lines of type can be arranged and locked. The labels may be fed as singles or from a reel and because of the quick set-up time even quite short runs are worthwhile. Labels may be of non-gummed, the traditional remoistening gum type, or the newer self-adhesive versions.

Work that is standard to size, *eg* paperbacks, may be more economically packed in cases than in parcels. These arrive flat, are erected *in situ* and sealed with gummed tape. High-quality commercial printing, *eg* die-stamped letter-heads and cards, may be boxed to prevent crushing that may occur in parcels. Books that are mailed direct to customers, *eg* book club work, mail order catalogues and prestige publications, often have specially designed cartons. These sometimes have built-in protective units that may include shaped expanded polystyrene blocks.

10. Standard mechanisms

Sheet gluing

The coating of sheet or real materials with wet adhesive is a process common to many specialist sections of the industry. Small quantities are readily dealt with by the usual brush method and this is really the best method of applying adhesives, the bristles working the paste or glue into the fibrous structure of the substrate, resulting in a first-class keying effect. Brushes used are usually of the round type and of a diameter suitable to the task in hand. Natural animal bristles and those deriving from man-made fibres are used, sometimes together in the same brush; the use of inferior quality brushes is usually a false economy as the bristles break down at a relatively fast rate. These are set into the shaped wooden handle and sometimes vulcanised; a tight metal binding prevents disintegration in use. A good brush will have fairly long bristles tied about halfway down with wire, which may be removed as wear occurs.

Most brushes require soaking before use and then should have the head always left moist to prevent the bristles being easily dislodged. Brushes should not be overcooked during use as this will result in the bristles breaking off in short lengths; if left standing on the bristle an unwanted curl will result; for some paste applications string bound brushes are used.

As an alternative method of applying adhesive by hand some companies use a hand-held roller; these are identical to paint rollers and by choosing a roller of correct width an even coating can be applied at high speed. This method is particularly useful for 'spot' or 'strip' gluing in the assembly of display material in small quantities and with the application of cold adhesives.

Sheet-gluing machines are designed to coat sheeted material prior to that material being further processed. 'Gluing' in this context means the application of any adhesive either hot or cold, and includes both bench and portable types with roller widths from 150mm upwards. When required the delivery of the glued material can be arranged wet side up or down and in some instances may be delivered direct on to a conveyor belt for transfer to the next part of the assembly process (fig. 10.1).

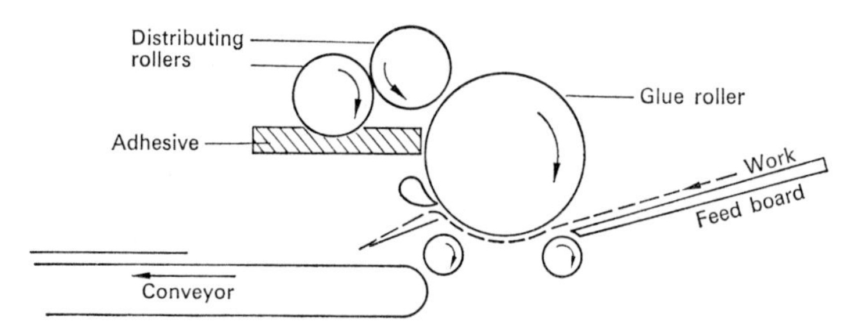

10.1 *Sheet gluing machine delivering side up to conveyor.*

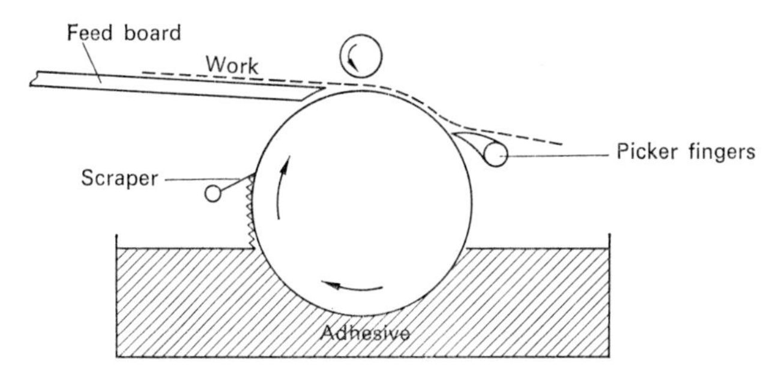

10.2 *Standard hand-fed and take-off sheet gluing machine.*

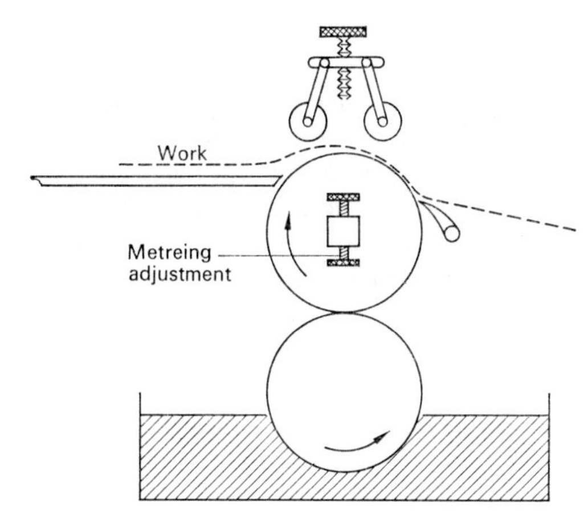

10.3 *Double-roller sheet gluing machine.*

The single roller type has a scraper or doctor blade to control the coating on the roller (fig. 10.2). The top roller of the double roller type is adjustable vertically and in this way coatings of different depth are obtained (fig. 10.3) and this type is often preferred because of superior distribution obtained. Under certain conditions of glue viscosity and roller setting, striations or ridges of adhesive appear on the applicator resulting in uneven coating.

For gluing rigid materials such as thick strawboard, it is necessary to prevent the sheet from following the contour of the roller. This is achieved by adding a false table to the machine, thus allowing the rigid material to pass across the top of the roller picking up the adhesive without distortion.

A very successful gluing machine is one that has a continuous wide rubber belt coated with adhesive. This machine may be fed in the usual way, but small area gluing is often effected by dabbing the material against the moving surface and picking off the glue at very high speeds (fig. 10.4).

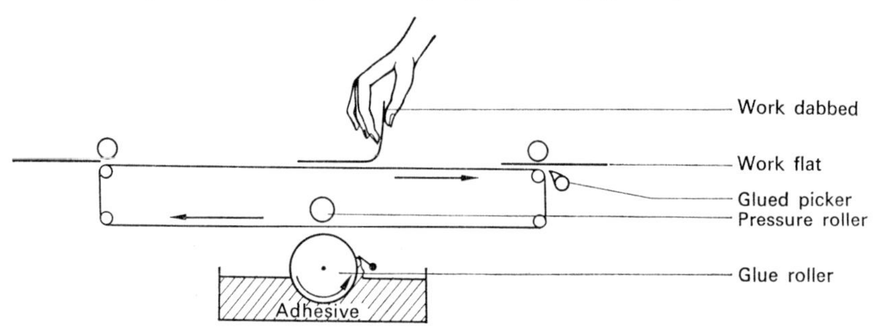

10.4 *Rubber blanket sheet gluer being used for 'dab' gluing.*

Casemaking machines

The cover-gluing mechanism on a sheet-fed casemaking machine is of a more sophisticated design. On this device the cloth cylinder and the glue roller are about 3mm apart and the cylinder only rotates when an actuating stud links it to the movement of the machine. The cloth cylinder is 'dressed' at the leading edge with a wedge-shaped dressing. This may be of the operators own making or may be a shaped plastic strip supplied by the machine manufacturer. When the cloth is laid on the cylinder, gripped and rotated, the thickest part of the dressing presses the leading edge of the cloth into contact with the glue roller. Because of the tacky nature of the adhesive the cloth is kept in contact with the glue roller although its face is away from the cloth cylinder by several millimetres (fig. 10.5). In this way the mechanism makes it possible to deliver the cloth to the assembly part of the machine with the face of the cloth clean and free of adhesives.

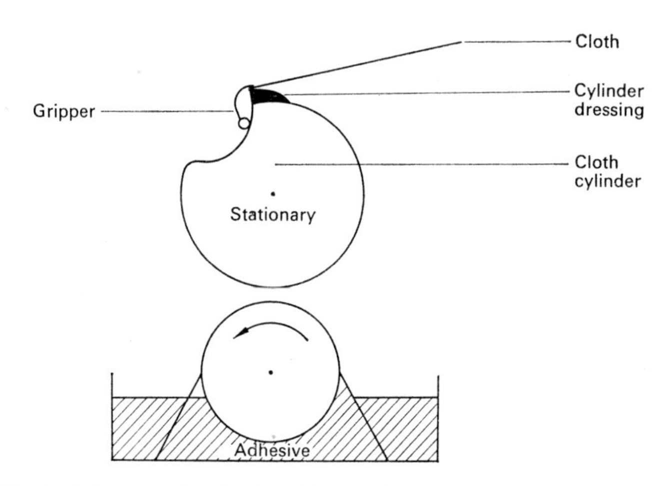

10.5 *Standard glue mechanism for sheet-fed casemaker.*

This glue roller rotates in both directions keeping the glue mixed and the levels on both sides of the roller equal, but of course requiring a scraper at both front and back to control the coverage in each direction (fig. 10.6). The newer twin-head sheet-fed casemaking machines have a larger glue tank with the glue roller rotating in one direction only.

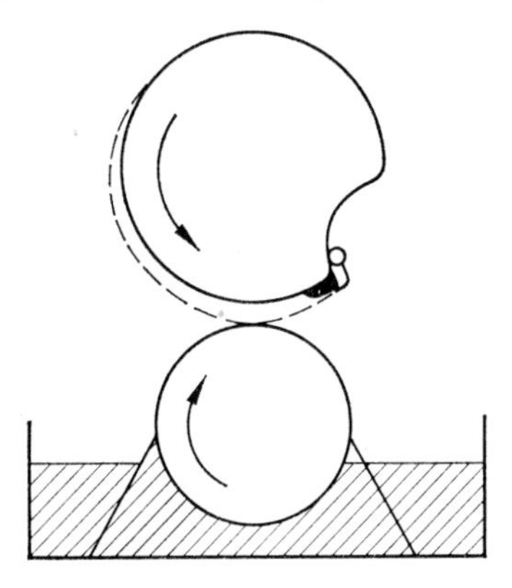

10.6 *Casemaker gluing mechanism in operation. Note dual scrapers and redirection roller.*

Stencil, spot and line gluing

There are very many instances in finishing operations where the application of adhesives needs to be confined to a specific area. Examples of this are:

1, gluing only the turn-in of a diary case to leave the case limp or soft sided; 2, gumming the edges of window stickers or in strips down a page of perforated cash receipts; 3, putting adhesives around the edges of clear foil or glassine that is to be attached to covers etc as windows; 4, spot gluing for the assembly of display items, calendars, sample sets and the like.

Stencil and spot gluing are both completed on machines that closely resemble simple printing devices. To the plate cylinder is fixed the shape of the area to be glued. This can be in rubber, plastic foam, cork or similar flexible materials. The sheet material is fed from a pile, automatically gripped on the impression cylinder and a print made in adhesive from the stencil. The image may be any shape to suit the job in hand and both hot and cold adhesives may be applied. Some glue tanks of sheet-fed casemaking machines may be replaced with stencil gluers, thus adapting the machine for the production of

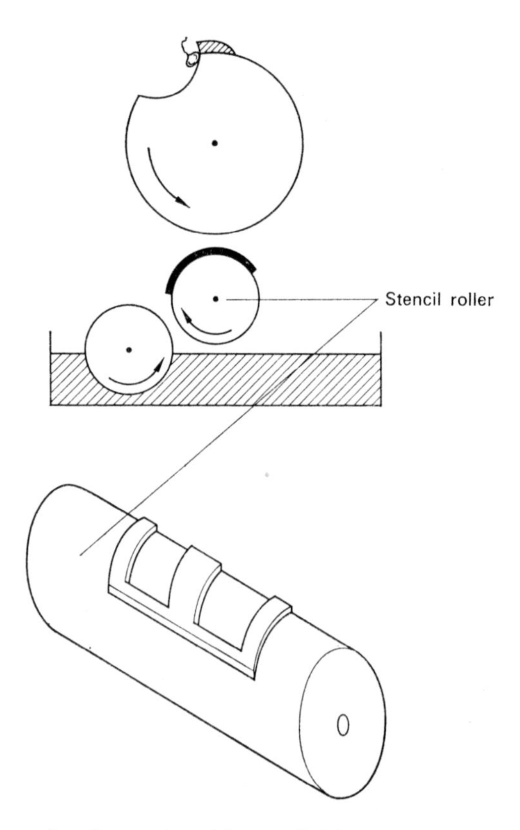

10.7 *Sheet-fed casemaker glue pot adapted for stencil gluing.*

satin- and silk-covered pads for greeting-card work, soft-sided book cases, etc (fig. 10.7).

The application of adhesives in strips or lines is done on the standard sheet gluer by replacing the main applicator roller with wheels of the required width. This is most useful when a strip of gum is required down a sheet of perforated receipts, etc.

Spine gluing

Spine-gluing mechanisms fall into two categories, those gluing sewn books and those for adhesive bound work. The requirements for a sewn book are slightly different as it is important that the adhesive penetrates between the sections and a minimum coating is visible on the spine after gluing. After application the adhesive is forced between the sections by a brush and the surplus taken off and returned to the pot; versions of this mechanism apply the adhesive from either a vertical or from a horizontal roller (fig. 10.8).

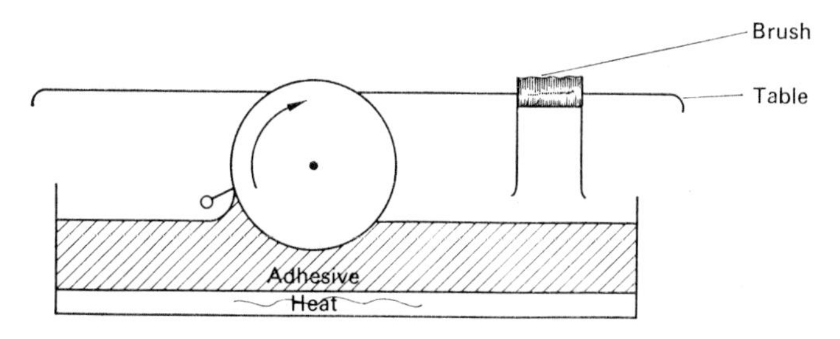

10.8 *Standard glue mechanism for spine gluing sewn books.*

Gluing adhesive bindings usually incorporates one glue roller with either doctor blade or roller to control and metre the film. In some instances a reverse roll mechanism is incorporated to both metre the film and to ensure accurate distribution of the adhesive on the applicator roller; double tanks and bars to remove unwanted strings of glue are also used. (fig. 10.9) A glue tank for gluing both the calico and the spine of the book of a calico-lined publisher's case book is illustrated in fig. 10.10. Another machine in the same

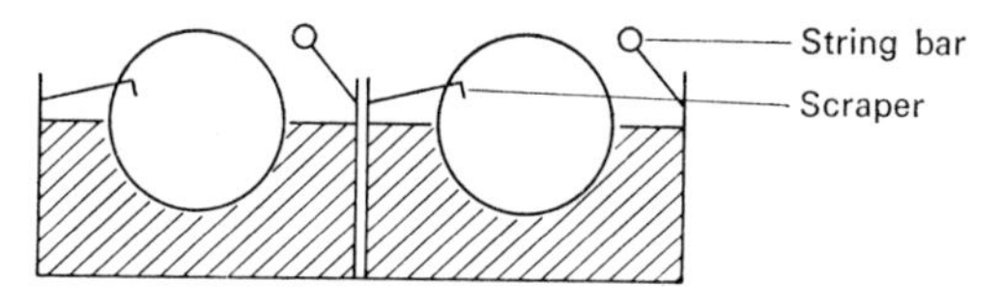

10.9 *Double glue tank for spine gluing adhesive bound work.*

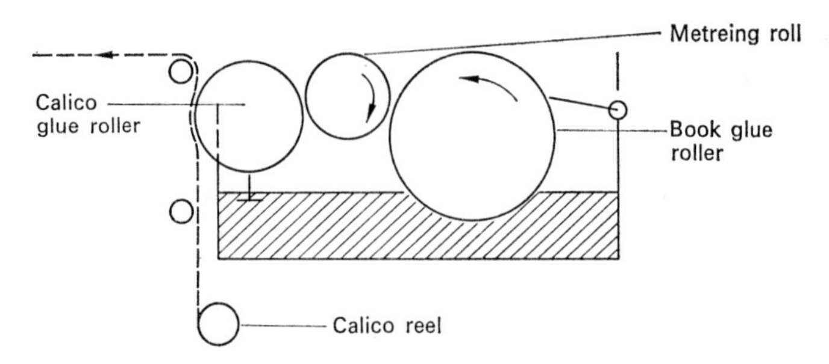

10.10 *Glue mechanism for adhesive lining machine. Roller for gluing the lining is friction-driven when gluing and power-driven when idling.*

field utilises a pressurised system with a series of orifices in a vertical tube. The unwanted orifices are masked out and the adhesive is pumped directly on to the spine of the volume (fig. 4.24). Surplus adhesive runs off and is collected in a pan ready for returning to the sump where it is filtered and reused.

New types of glue pots for adhesive binders are currently undergoing field trials. One pot utilises the gravure printing technique, the adhesive being applied in the typical gravure configuration. Another hot melt applicator has a cam-controlled doctor blade. This determines both the film thickness and length of the roller covered so that head and tail of the spine may be left unglued. As this area coincides with the trim line the offcuts will not contain any hot melt, thus ensuring highest price for trimmed waste (fig. 10.11).

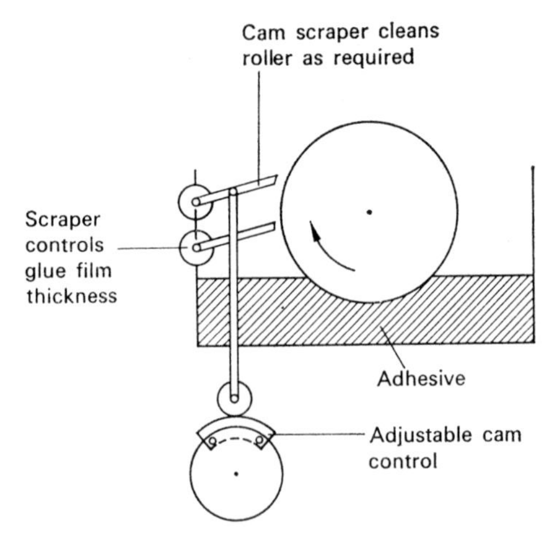

10.11 *Cam-controlled doctor blade on glue tank of adhesive binding machine.*

Pasting endpapers

Application of the relatively slow moving pastes and starch/PVA formulations on to endpapers during the casing-in process has one main complication, *ie* the necessity for ensuring that sufficient adhesive finds its way into the area below the shoulder. The simplest is one that will paste the endpapers of books that have not been 'backed'. A modification to accommodate the shoulder is shown in fig. 10.12 and although the applicator roller is of larger diameter than in the previous example, the maximum length of book that can be pasted is governed by the circumference of the roller.

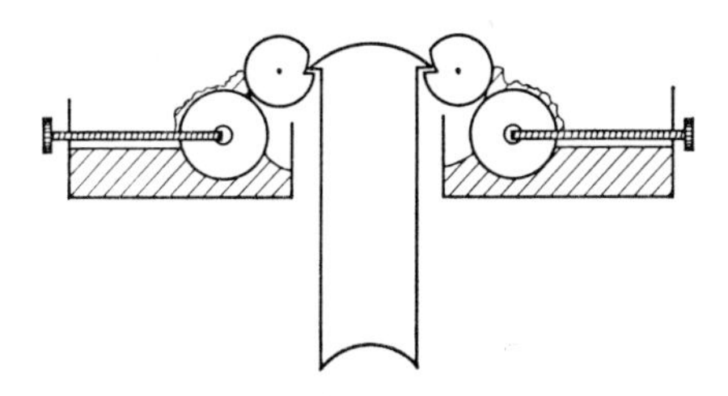

10.12 *Paste box incorporating grooved rollers.*

A somewhat more sophisticated approach is contained in paste boxes that incorporate the half reverse-turn mechanism. Here the applicator roller is of relatively small diameter giving a much better fit into the angle of the shoulder. Before making contact with the endpaper the rollers are reversed by half a turn. The upper scraper allows a thicker application to the roller so that, on contacting the mull and endpaper, extra adhesive is squeezed into the angle. The roller now rolls over the endpaper, the paste carried being controlled by the bottom scraper bar (fig. 10.13).

To ensure that joint gluing in fast-running machines is positive and to overcome certain problems associated with the nature of some lining

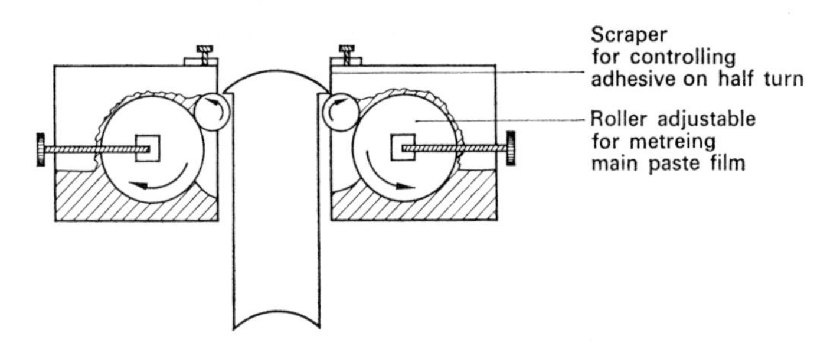

10.13 *Paste boxes incorporating the half turn mechanism.*

materials, a new device has appeared that applies a narrow strip of PVA to the endpaper just below the shoulder (fig. 10.14). There is also some advantage in the application of PVA formulations at this stage in the production sequence as it can be dried quite rapidly by the creasing irons in the book forwarding and pressing machines.

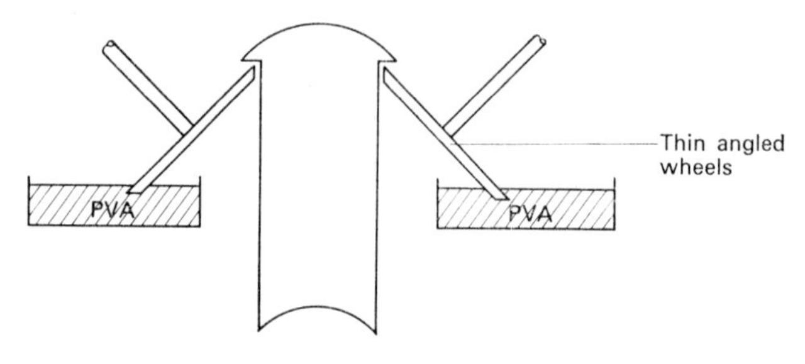

10.14 *Joint gluing attachment.*

Strip pasting

Attaching plates and endpapers and pasting sections together during sewing are examples of applying adhesive in narrow strips.

The adhesive on modern endpapering machines is usually applied by a simple narrow wheel that picks up the adhesive from a reservoir of paste. Other types utilise a bar which carries paste from a duct and roller on to the edge of the section. A similar device is used on sewing machines to paste the first and last two sections together.

Nozzle applicators

A method of applying hot melt adhesive in strip form uses a nozzle applicator. The hand-held version is loaded with plugs of the adhesive which is heated to the required temperature and ejected by a trigger movement of the control.

A larger, machine mounted version, is fed from a continuous 'rope' of adhesive which, after melting, is ejected upon instructions from the machine. The obvious advantages of this method are: 1, only a small amount of hot melt is being maintained at a high temperature with resultant savings in heating costs; 2, the adhesive is wholly enclosed preventing loss of heat and fumes; 3, greater control can be achieved in the application of the adhesives. Main use of these applicators is at present in the carton gluing and sealing field.

An applicator that may be fitted to some folding machines glues the centre fold of an 8-page or 12-page section to the outside fold and thereby makes wirestitching unnecessary. The mechanism consists of an adhesive pump applicator, two sensing devices, adhesive container, pump unit and control box. Pressure supplied by the pump unit at about 2·8 kgf/cm^2 is reduced to 0·5 kgf/cm^2 by a control valve and this pressure is used to convey the adhesive from the container to the pump applicator. The light-operated sensing device reads the leading edge of the incoming sheet and triggers the solenoid-operated valve in the pump allowing the adhesive to flow under pressure. Similarly the trailing edge sensor closes the valve.

The applicator nozzle has an aperture of 0·4mm and is set to just brush the paper surface flowing beneath for about three quarters of the length of the page; gluing right up to the trim line is undesirable. Grooves cut into the first fold rollers prevent the adhesive being picked up at this point. Special sensors may be needed for printed work containing many solid tints.

Section feeders

The automatic selection of a single section from a pile and placing it in a position where it can be transported for further processing tends to be an awkward task. This type of feeder is appearing more frequently than in the past as manufacturers attempt to remove the hand element in machine production. These devices sometimes put up machine running speeds, but when they function well always relieve the operative of tedious hand laying-on and allow a regular uninterrupted flow of work through the machine resulting in higher net production figures.

Two different types of feeders are in use: one that simply removes a section from the pile and lays it on to a transporter and another which selects the section and opens it to centre before laying on. All feeders have air suction pads to draw down the section from the pile, the stack being kept in place by retaining fingers. The power of the pump employed must be sufficiently great

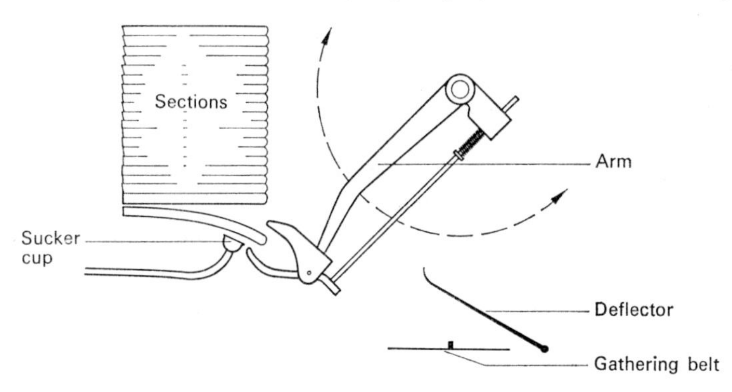

10.15 *Principle of reciprocating gathering arm.*

to draw down all weights of section likely to be handled and this tends to be one of the limiting factors of machines.

Traditional gathering machines utilise a reciprocating arm and gripper to grasp the section, withdraw from the pile and deliver to the transporter. The arm embodies a pre-set calipering mechanism which only functions if the correct amount of paper lies between the fingers; doubles and misses cause the trip mechanism to operate and the machine to stop (fig. 10.15). Rotary equivalents are to be found on both gathering and endpapering machines. Here a rotary drum mounted below the pile of sections has a gripper mechanism reminiscent of that found on sheet-fed cylinder printing presses. As the gripper passes the leading edge of the section it closes and holds it tightly to the cylinder withdrawing the section from the pile. At the bottom dead centre point the gripper opens allowing the deflector to direct the section on to the transporter below (fig. 10.16).

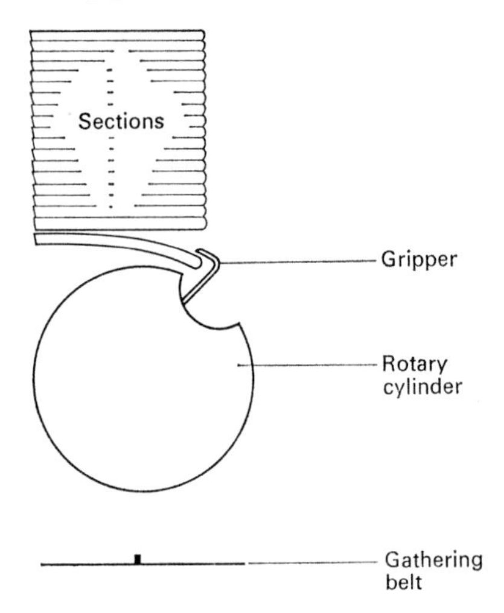

10.16 *Principle of rotary gathering head.*

The feeding and opening of sections that have to be placed over a saddle-type transporter chain is a much more complex operation and the problem is approached in one of two ways depending upon the method of folding used and the make-up of the sections concerned. The mechanism has an initial rotary-feed cylinder that selects the section from the pile in a similar manner to that previously described, except that it is carried about 240° around the periphery before being released. By this time subsidiary opening drums have located the fore-edges of the sheet and pulled them apart. A separate placing

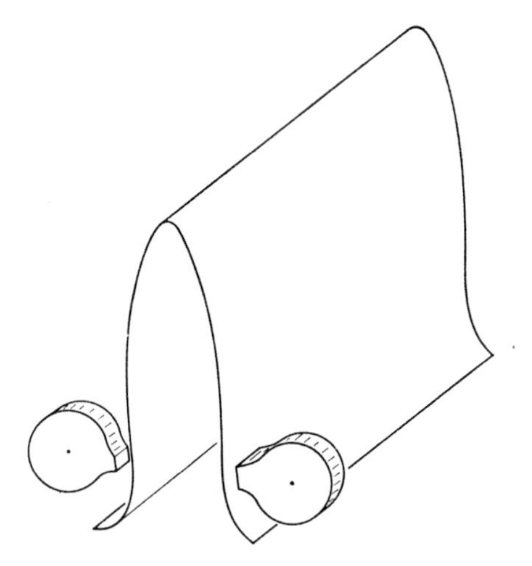

10.17 *Suckers open the section to the centre.*

mechanism then ensures their correct position on the saddle. Work that has a firm head 'bolt' (fold) and without any added units will be opened by suction pads embodied in the secondary cylinders (fig. 10.17). Other work having insets, wraprounds, tipped plates and endpapers has a number of false centres

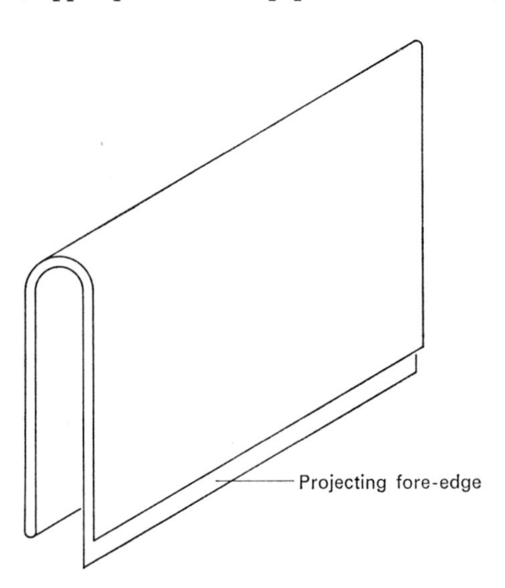

10.18.1 *Section with projecting fore-edge suitable for feeding on automatic section feeding mechanisms.*

and cannot be accurately opened with sucker mechanism. These sections have to be folded and made up so that one half of the total thickness has a 'projecting' bolt. This is an extra area of paper approximately 5mm wide which gives the appearance of the section having been folded off centre (fig. 10.18.1). All the subsidiary parts of the section are similarly folded so that after make-up all the projecting bolts are to one side. The subsidiary cylinders are now fitted with mechanical grippers that can locate the projecting bolt and find true centre (fig. 10.18.2).

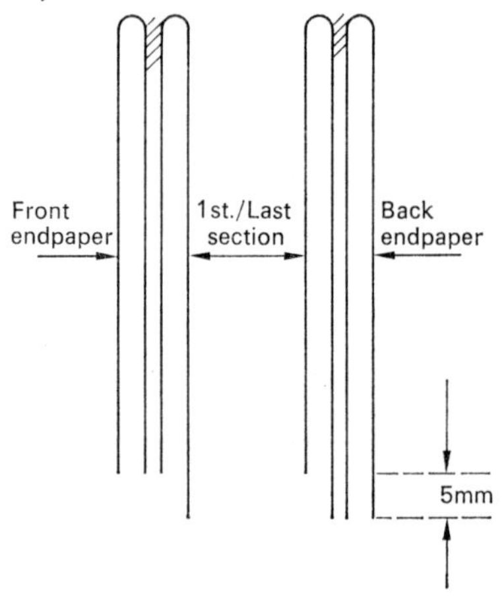

10.18.2 *Arrangement of endpapers for gripper opening on automatic sewing machine.*

There are certain inherent difficulties in organising the printed work so that this equipment may be used. Much of the inset magazine work is fairly simple in make-up and the time between imposition and binding relatively short. With the fairly long runs involved it becomes an economic proposition to impose the work specifically for the automatic insetter-stitcher to ensure that it comes within the requirements of the feeder and this is reflected in the large number of installations now using this equipment.

In the publisher's case binding field feeders of this type may be used in book sewing but the situation is complicated by a number of factors. The composition of sections within a book is not necessarily standard and may contain various mixtures of 32-, 16-, 8- and 4-page units either right angle or parallel folded and with 2-page illustrations and endpapers tipped on. Although it is theoretically possible to organise the make-up to suit the auto-feeder it often is not justified on the short runs that are common in this type of work. Another complicating factor is that a sizeable proportion of publisher's case work is produced in specialist trade binderies who take in printed matter

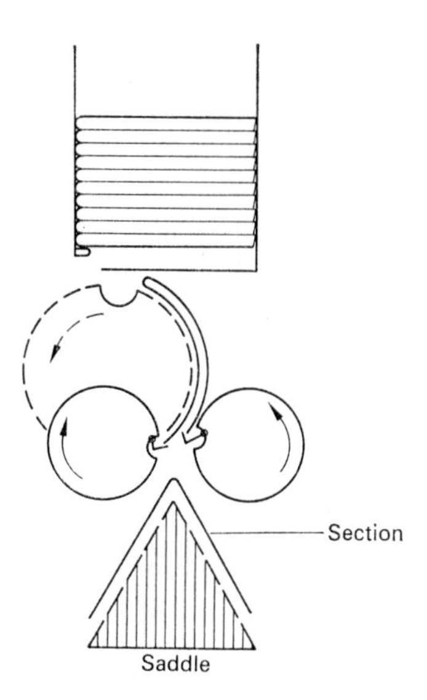

10.19 *After the main gripper cylinder has extracted the section from the hopper, subsidiary (or planetry) cylinders engage the fore-edge projection and lay the open section on to the saddle.*

from a wide variety of printers and often at the time of printing the destination of the sheets is unknown. With these difficult organisational problems to solve and the high outputs achieved by hand-fed semi-automatic sewers most companies are reluctant to forfeit flexibility for a relatively small increase in output at high capital cost. Automatic feeders are, however, used in the large printer–binders where the product is of straightforward make-up and in large quantities, *eg* mail order catalogues (fig. 10.19).

Pressing

Pressing is necessary at various stages of print finishing to consolidate, ensure adhesion, reduce bulk or simply as a clamp to hold the unit firm whilst being processed. Wooden screw presses are still favoured by hand craftsmen for lightness and the general handling qualities of the material; it is sympathetic to hand work and will not rust or otherwise mark the delicate leathers that may be pressed. With care wooden presses have a very long life and both laying and finishing presses are still available.

Vertically operating hand 'nipping' presses are usually all steel or cast iron in construction and make use of the powerful square thread; power is applied through a weighted swing lever or wheel. 'Standing' presses are four pillar presses which are floor mounted; pressure may be applied through a wheel or a long pin or lever. Both types are to be found in regular use in the hand binderies but while exerting as much pressure as needed with quite a small effort they are time consuming in use. A simple press that combines screw and lever pressure is a 'set limit' version; the platen is set to a given height through a screw and the bed is then lifted and pressure applied by a lever system. This press is very useful when piles of identical height are being continuously processed.

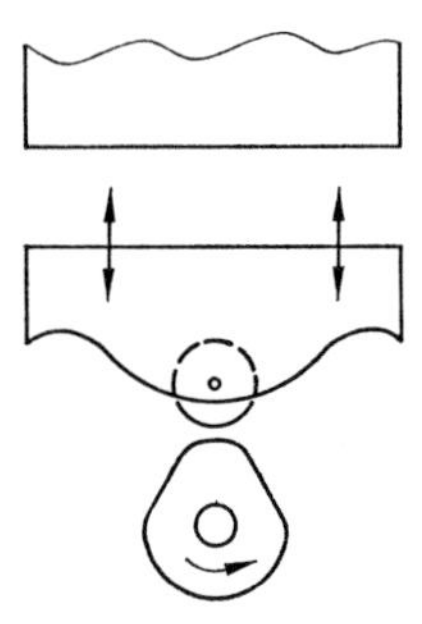

10.20.1 *Cam-operated press.*

Mechanically powered presses perform many functions in mass production work and utilise the full range of mechanical, electro-mechanical, pneumatic and hydraulic principles. Mechanical principles used include heavy cams, toggle joints and eccentric reciprocating beams (fig. 10.20.1, 2.3). These are to be found on nipping, blocking and punching machines of various power using small powered motors suitably geared down. Because of its action the toggle principle allows a large vertical movement at relatively high velocity and this makes it suitable when high speeds and great impressional power are needed.

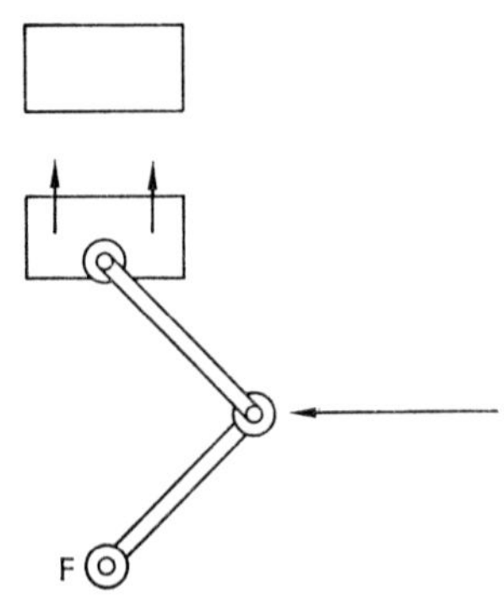

10.20.2 *Toggle-operated press.*

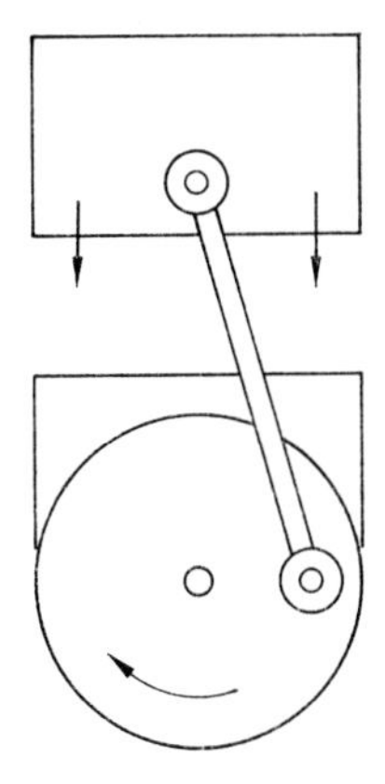

10.20.3 *Eccentric reciprocating beam press.*

One simple bundling press has an electric motor drive through an epycyclic gear train giving a ratio of 160:1. This supplies good compressive power whilst the mechanism is protected by a clutch which slips at a predetermined level. A similar principle is used in a small nipping machine when a brake band slips if the machine is overloaded.

Increasing application is now made of pneumatic and hydraulic power in pressing and clamping systems. Examples of pneumatic application are to be found in blocking as this facilitates high speeds and the ability to apply equal impressions into materials of variable thickness such as leather. A control valve can be set to blow off at a given pressure and the blocking head will 'feel' its way into the material until that pressure has been reached; the control will then function and the pressure will come off.

Pneumatics are also used for the application of pressure on the rollers of a gilding machine. An air cylinder located at the centre point of the heat transfer rollers apply up to 650 kN/m^2. This gives a very simple construction with great flexibility and easy pressure change for different types of paper. About the same pressure is used for pneumatic nipping and standing presses; the cylinder is located above the press platen and transfers around 10,300 kN/m^2 to the books. A three-way control valve is used having positions for down, hold and up (fig. 10.21).

Another very versatile tool using pneumatics can be adapted for punching, indexing and similar tasks. The head is powered by two air cylinders and actuated by a foot control bar. With careful jigging and gauging a wide variety of shaping and pressing operations can be performed. Other uses for pneumatics in print finishing include hand-operated indexing guns, bundling presses, control and redirectional mechanisms in automatic lines, press and feeder operation in high frequency welding.

The application of hydraulic power has now ceased to imply the use of water under pressure and is always concerned with oil as the fluid. Because theoretically oil is incompressible immediate and tremendous pressure can be imparted. This is useful when greater pressures than those obtainable with pneumatics are required and at high speeds. The engineering related to the application of hydraulics tends to be somewhat simpler than similar mechanical devices and lend themselves to flexibility of design. On one nipping machine up to 108kN can be brought to bear on the spines of books, without any danger from fracture of metal parts due to the insertion of an oversize book.

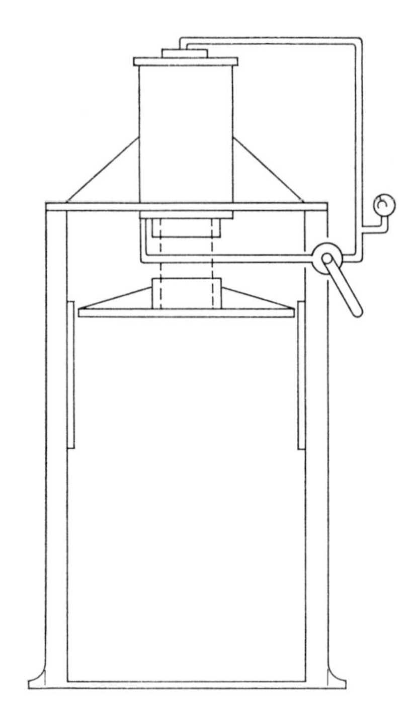

10.21 *Pneumatic standing press.*

Conveyors

The automatic transporting of goods or partial assemblies around the workshop may be both convenient and productive. Perhaps the simplest of these is the gravity roller which is placed so that a slight decline forces the object laid upon the rollers to move in the desired direction, the speed of movement depending upon the angle of the rollers and the weight of the object. These are used for on and off loading small parcels into road transport and at various points around the workshop to move work in progress. A typical application is conveying quantities of sections from a folding machine line to a bundling process area. By mounting two sets of rollers vertically and angled in opposite directions the bottom rollers will deliver the filled containers to the bundling area and when empty the containers are placed on the upper, return line (fig. 10.22). Similar lines are used for delivering goods to packaging areas and have the merit of rigidity, strength and random movement; no particular time cycle having to be observed. Powered versions and a variety of different designs are available.

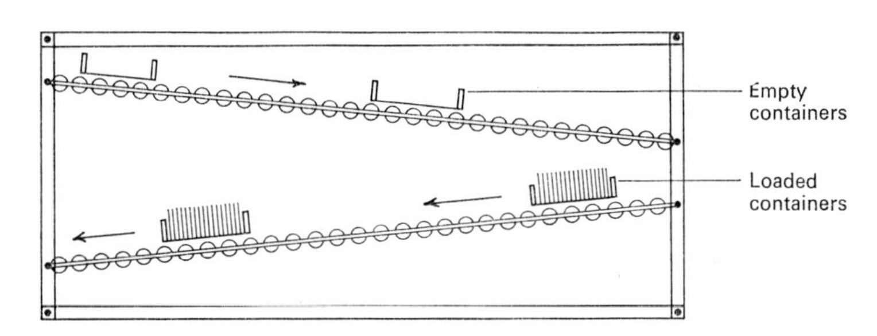

10.22 *Example of steel roller conveyor used for the random transporting of folded sections to a bundling area.*

Power operated conveyor belts are finding increasing popularity for light assembly work. Instances of their use are from the take-off position of a line of sewing machines to a central inspection and stacking area; and a hand

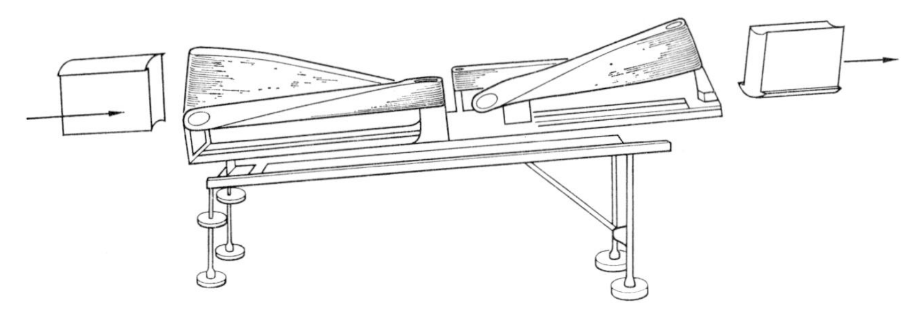

10.23 *Specialist belt conveyor for inverting and transporting books between in-line machines.*

mailing line that necessitates assembly of several items in a box before labelling, franking and posting. Versions are available that can be operator started and stopped and these are useful to serve a number of operatives with material from a common point, *eg* glued sheets of cover paper and liners for the manufacture of calendar boards. The use of specialised conveyors has made the automatic binding of books possible in recent years. These rather narrow belts collect books from one machine and deliver them to the next sometimes turning the book through 180° *en route*. This type usually terminates in a mechanical timing mechanism that controls the infeed of books into the following machine (fig. 10.23).

Purpose-built conveyors include a wide variety of designs to transport small quantities of goods at a time over quite large distances. Examples of these can be seen in large installations picking up from a printing press or binding line and delivering to a publishing or mailing room by overhead means.

11. Modern guillotines and folding machines

Development of the guillotine is, perhaps, progressing more rapidly than any other single piece of finishing equipment, and machines are being installed to meet the demand for greater accuracy and higher outputs. One aspect of machine design that is also developing is that of safety, both for the operator and for the machine. The only types of guards acceptable in the United Kingdom for single-knife paper-cutting machines are 1, automatic sweep-away guards; 2, automatic push-body guards; 3, interlocked guards; and 4, photo-electric devices.

The types 1 and 2 are so arranged that the operator is thrust away from the cutting area very soon after the knife movement has commenced. With the guard fully extended it is impossible to reach the knife, the distance being too long for the human arm. The term 'interlocked' implies that the guard or screen must be in its closed position before the knife can be set in motion and in some instances opening the guard during knife motion will bring the knife to a halt.

Photo-electric devices are advantageous as they remove the clutter of bars and brackets, associated with the previous types, away from the working area leaving the worktable free for the manoeuvring of the stock; any action that breaks the beam will halt the knife. Many modern machines now have double push button electrical starting systems but these have to be incorporated with any of the previously mentioned guards to be acceptable.

Important safety features are provided when magnetic clutch/brake mechanisms are fitted as these automatically go to 'brake' when stopped or in event of a power failure. Safety shear pins and burst washers are incorporated to prevent damage to the machine by mechanical fault or overloading.

Springs, friction and hydraulic pressure are used in the clamping systems; the advantages claimed for hydraulics are greater clamping power and ease of adjustment. Typical maximum clamp pressure provided by springs is about 24·5 N on the clamp area while with hydraulics the clamp pressure is infinitely variable up to a maximum of 39·0 N. A balanced pressure cylinder enables the clamp to be treadle operated for sighting with a pressure of 0·50 N (fig. 11.1.1).

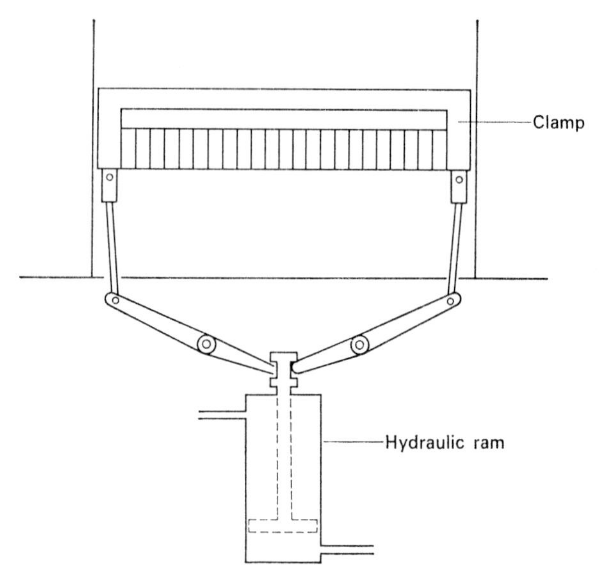

11.1.1 *Hydraulic guillotine clamping system.*

The improved electro-magnetic clutch mechanism also gives better transfer of power to the knife thus eliminating drag and other cutting faults. The combination of dip-shear action and double-end pull distributes the load, spreads the reacting forces and produces a uniform action (fig. 11.1.2).

The back gauge may be fitted with devices for angling horizontally or vertically to overcome faults in cutting difficult stock or inaccurate printing,

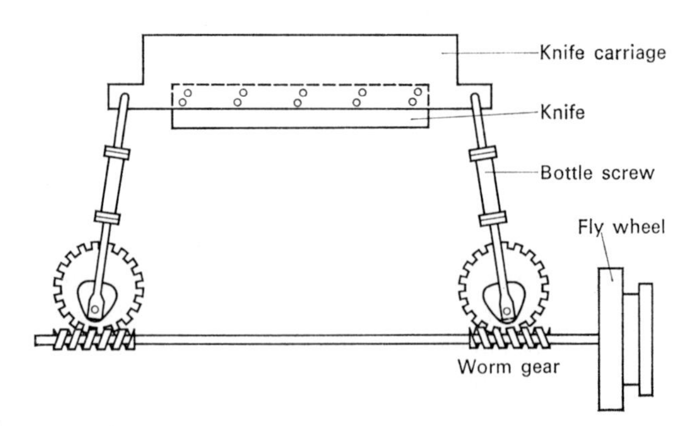

11.1.2 *Double-end pull-on knife beam.*

eg all sheets identical but out of square. The gearing used to move the back gauge has been improved so that an accuracy of 0·05 mm can be obtained.

Other devices that appear on guillotines and help to improve output include airflow tables and vibrators, improved dimension-reading devices and newer guards that leave the work table completely free. Much of the time of a guillotine operator is spent rotating reams of paper horizontally on the work table to enable the various edges to be trimmed. This is particularly so in mills and those houses where accurately trimmed reams of white paper are necessary in the printing process. The airflow table is used to combat the tiring and slowing effect this may have by supporting the bulk of the weight upon a cushion of air. The work table of the guillotine has a series of small holes drilled into it and each is connected to a blast pump supplying air and forcing small balls into the sockets. When the ream of paper is laid on to the work table the balls are depressed by the weight of the ream, against the air pressure and the ream is partially supported thus reducing the effort needed to rotate the pile (fig. 11.2). Well lit, magnified tape reading devices are appearing that have a display panel showing figures, the units and decimal fractions on which the back gauge is set.

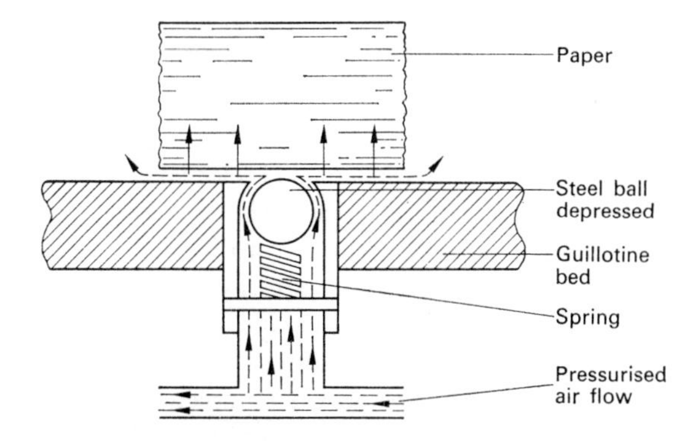

11.2 *Principle of airflow table showing one outlet in use.*

High speed cutting usually has very little to do with the number of cycles per minute that the knife can operate as most of the time is spent manipulating the paper or resetting the back gauge and it is this last factor which has received most attention. The speed at which the back guage will run out to maximum cut and then run in is very important; machines with power operated back gauges often have high speeds for the backward motion and slow speed forward with a creep speed for the last 50 mm. This is necessary if the stack of paper is to be brought to a halt without the displacement that would occur if a high forward velocity was used.

Means are provided to advance the back gauge in preset dimensions by a hand-operated mechanical device but usually only one dimension can be set. A more sophisticated electro-mechanical method has a sliding electrical switch that is connected to the movement of the back gauge. Flags or stops are placed along a rail in the required positions and as the switch passes the flag a contact is broken bringing the switch and back gauge to a halt. A number of rails are available making a quick changeover and storing of master rails possible; while being relatively crude this system has the advantage of the programme being entirely visible to the operator. An improved version uses a photo-electric cell in place of the micro-switch and this allows finer cutting distances.

The application of electronic controls to single-knife guillotines has revolutionised their appearance, versatility and outputs. Basically the system adopted has been the use of sensitised magnetic tapes with suitable marking and reading heads working in conjunction with the back gauge. The magnetic tape may be attached to a cylindrical metal drum or flat steel surface to prevent stretch occurring and up to 64 parallel channels may be available. The marking and reading heads will be stationary above the rotary drum but will slide along in the case of the flat tape and will be so controlled that the operator can mark or read any of the channels available by pressing the appropriate button.

To set a programme a particular channel is selected and the back gauge is run out to maximum, usually at high speed. The back gauge is then brought forward to the required dimension, accurately positioned with the mechanical micrometer setting and the marking button pressed. This marks the tape with an impulse dot and will be available on the tape until it is no longer required and is wiped off electrically. Marking continues on the forward movement of the back gauge until the cutting pattern is complete; after switching from marking to reading a proving run is made, the stock placed in the machine and cutting begun. Both automatic and semi-automatic actions are usually available; in the latter case the back gauge will move forward under the control of the tape but operation of the clamp and knife requires action by the operator. This control is suitable for situations where difficult work is being cut and checking or ream manipulation is required between cuts. By switching the machine to 'automatic' the knife action is integrated with the back gauge movement so that the knife descends automatically after each forward movement of the back gauge. Most reading heads have a forward scanner that anticipates the stopping of the back gauge and slows the motor so that during the last 50mm or so of travel the back gauge is moving relatively slowly. This prevents jerking of the pile as the back gauge stops and allows many classes of paper to be cut automatically (fig. 11.3).

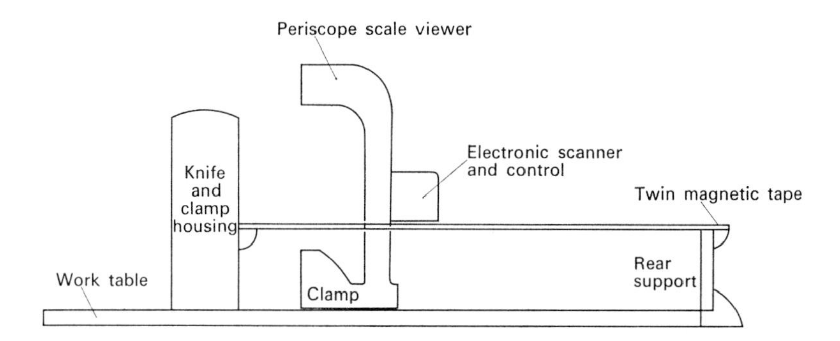

11.3 *Diagrammatic side view of a guillotine having back gauge under magnetic tape and electronic control.*

It will be noted that the scanning or reading by the head only occurs during the forward motion of the back gauge. If the cutting programme requires a long cut to follow a shorter one, *eg* if, in squaring reams a long edge is trimmed followed by a shorter edge, then it is necessary to mark two channels and link them. This may be completed automatically and some machines can link up to four programmes in this way making a very complicated and comprehensive cutting sequence possible. These machines are relatively fast to set up and on some machines facilities for permanent storage of cutting patterns are possible by removing the magnetic tape and its holder into storage. Spacing of cuts as close as 1mm may be possible and allied to the accuracy standards previously mentioned very close tolerances can be met. Disadvantages of sophisticated machines of this type, capital cost apart, are shut-downs due to electronic faults and ghosting as the magnetic tape wears; when on long runs, a check on dimensions should be made daily.

The latest advance in this field is the application of the digital computer to the programming facility. In this case the computer monitors the lead screw operating the back gauge and will stop when so instructed. To cut labels 120mm × 80mm printed with 5mm trim requires that two linked programmes be used; first setting is 120mm and 5mm linked for alternate cutting, the back gauge moving forward these distances alternatively until complete. The settings are then erased and reprogrammed for 80 and 5mm. Although it is necessary to re-dial for every set of cuts this is quickly accomplished and abolishes the storage of tapes and bars. Accuracy standards are high but application is restricted to highly specialised work with long repetitive runs.

Although many guillotines are used as individual machines greatly increased productivity can, in certain circumstances, be achieved if it forms part of a more comprehensive productive system, and is well backed up by handling and other devices. In the case of a machine confined to ream trimming, the physical task of lifting the paper on to the machine and then lifting it back to the pallet consumes most of the productive time. One approach to this problem is the provision of hydraulic lifting rams set either side of the machine; one lifting the incomplete stack so that it is constantly level with the work table and the other taking the completed work and moving down so that it too maintains the correct working height (fig. 11.4).

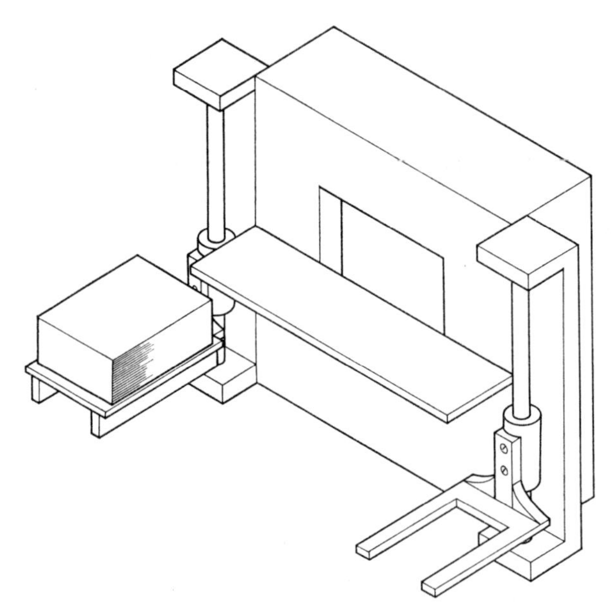

11.4 *Synchronised lifting rams as part of the guillotine equipment.*

A system for reducing paper into cut sizes is illustrated in figure 11.5. The work surfaces surrounding the guillotines are all air-flow, enabling the work to be moved forward with the minimum effort. In some instances large users of paper find side-loading, *ie* directly on to the back work table, and the provision of joggers and specially engineered in-feed mechanisms in the line a highly productive system.

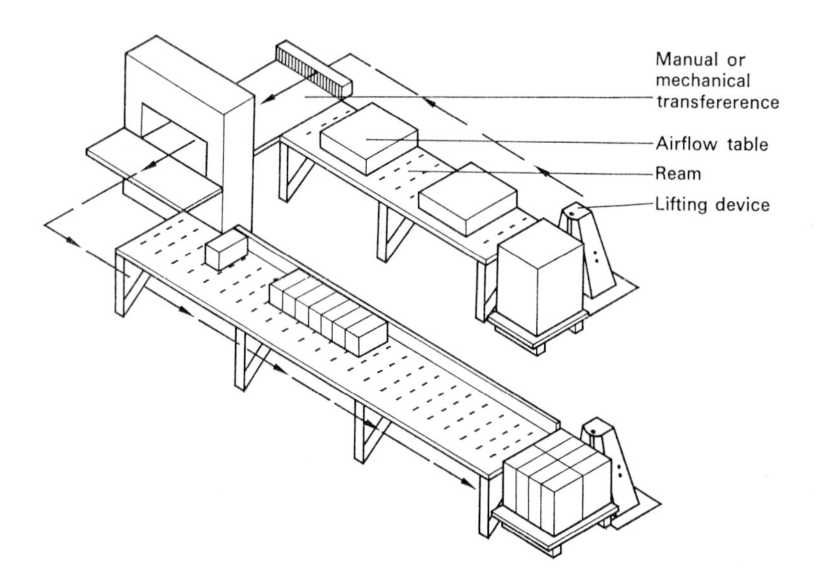

11.5 *A system for the production of cut work using a side or back loaded guillotine.*

Multi-knife trimmers

The high speed trimming of all classes of work during or at the completion of the binding process is one of great importance. If this is the last task in the production line, then a clean square professional finish will enhance the product. If further processes are necessary before completion then accuracy is essential as the book block will certainly have to marry up with other units. A number of basic types are available.

Pile trimmers

The slowest but perhaps the most flexible in application is the 'striker' trimmer, in which the work is loaded into the cutting position by hand, clamped by foot operated mechanism and the trimming cycle started by the operator. This type of machine is relatively cheap and lends itself to difficult jobs and short runs.

Semi-automatic in-line pile trimmers progress the work in a straight feed-cut-deliver line. The pile is placed in the in-feed clamp which grips and carries it into the cutting station. The book clamp lowers into position before the in-feed clamp releases so that the pile is at all times under control (fig. 11.6). The timing is such that the trim cycle is already under way as the in-feed

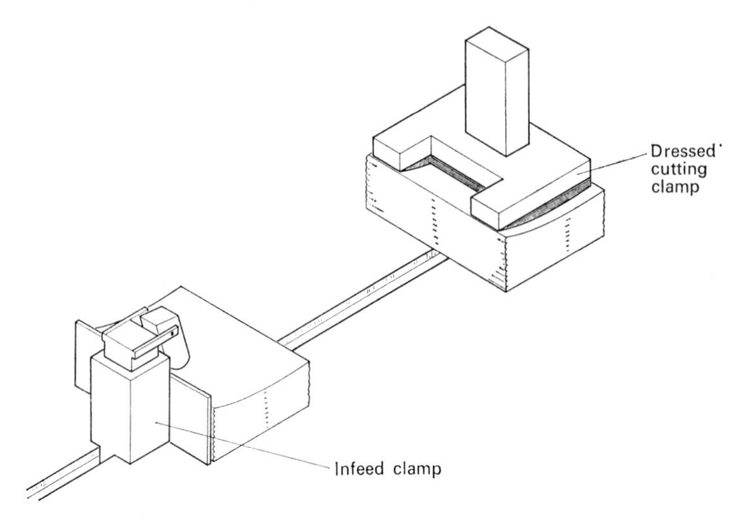

11.6 *Infeed clamp and dressed book clamp on three-knife trimmer.*

clamp starts its return journey to be reloaded. The head and tail knives trim first, moving through an oblique path and then retracting high enough to clear the fore-edge knife. As the knives reach their neutral position the book clamp pressure comes off and the action comes to a halt (fig. 11.7). The second pile fed-in may start a book delivery mechanism that pushes out the first pile, or the fore-edge of the incoming pile may perform this function.

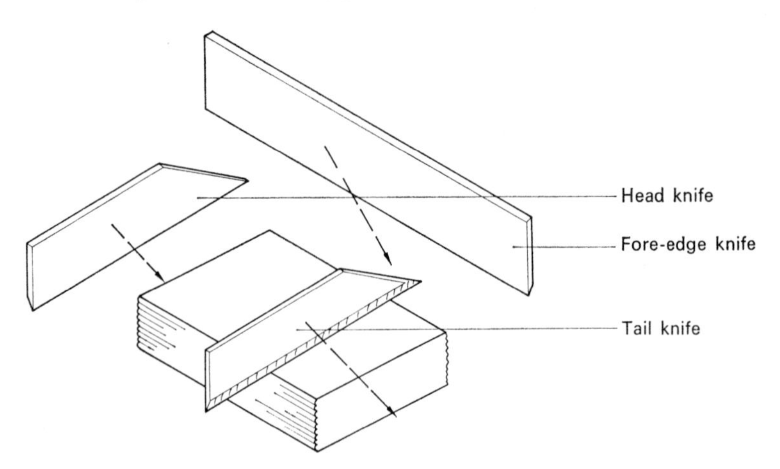

11.7 *Relative position of knives on three-knife trimmer.*

Although the piles are hand positioned by the operator the in-feed clamp may be either continuous or triggered off by the operator according to the class of work being handled. Piles are limited in height to the amount that can be held by the human hand and around 100mm may be considered a maximum; machine running speeds are up to 25 cycles per minute. Most machines of this type are hand fed but a continuous feeder is available forwarding the piles from a hopper or previous batch counter. The in-feed clamp of an in-line trimmer is usually arranged to grip the pile away from the swell in the spine and when positioned under the book clamp it, too, must be shaped to accommodate the swell (fig. 11.6). To help with feeding problems machines are equipped with variable speeds.

Another feature of this type of trimmer is the ability to divide two-up work, the tail half being returned to the operator for trimming during a second cycle of the machine.

Conveyor-fed straight-line trimmers utilise several trimming positions to achieve high operational speeds of up to 40 cycles per minute and because the pile is machine fed and not hand placed the piles can be up to 150mm high. Chain and peg conveyors forward the work along an inclined trough and pincer action grippers transfer the work from station to station where it is progressively clamped and trimmed. Both three- and five-knife versions are available and a typical layout is illustrated in figure 11.8 which shows a five-knife machine suitable for two-up work. The first knife trims the fore-edge; at the second station three parallel knives trim head and tail and split the books apart. At the third station the leading book pile is trimmed to remove

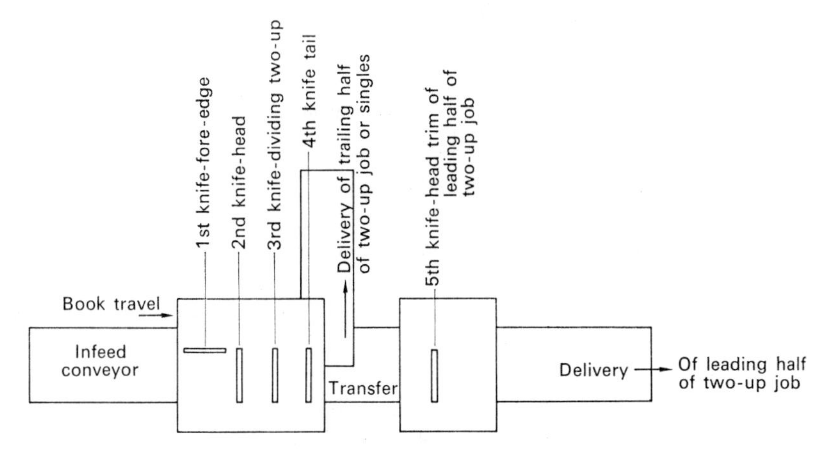

11.8 *Plan of five-knife pile trimmer.*

the bevel edge to finished size. Meanwhile the trailing half has been delivered to a belt conveyor.

On this machine cover break of thick square back work is prevented by a small groove milled across the back of the pile at the point of the fourth and fifth knife cuts. This type of equipment can be built into automatic lines producing both square back and saddle-stitched work and is particularly useful for trimming magazines which are folded and stitched on reel-fed printing machines.

Single copy trimmers

The trimming of inset and saddle-stitched work can be completed at the conclusion of the binding machine by a trimmer linked to and working at the same speed as the insetter-stitcher. As the work to be trimmed is usually not more than about 10mm thick a shear action knife is often employed. This has the usual top knife but the bottom cutting surface is a hard plastic, nylon or steel knife and the top knife shears past rather than into it (fig. 11.9).

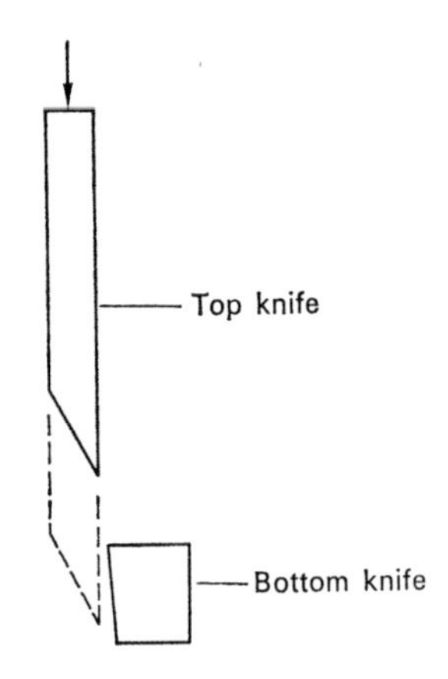

11.9 *Shear action used on some single copy trimmers.*

The book progresses along a pegged conveyor chain, spine foremost, and is transferred into the first cutting position and up to lays where the fore-edge cut is made. The book is then transferred to a second position by top and bottom friction tapes where it is registered and spring clamped before the head and tail knife descend (fig. 11.10).

Speeds for this type of trimmer are high and round 7500 copies per hour are common. Where two-up work is concerned models with a further one or two knives are available. To meet the demand of insetter-stitchers working at

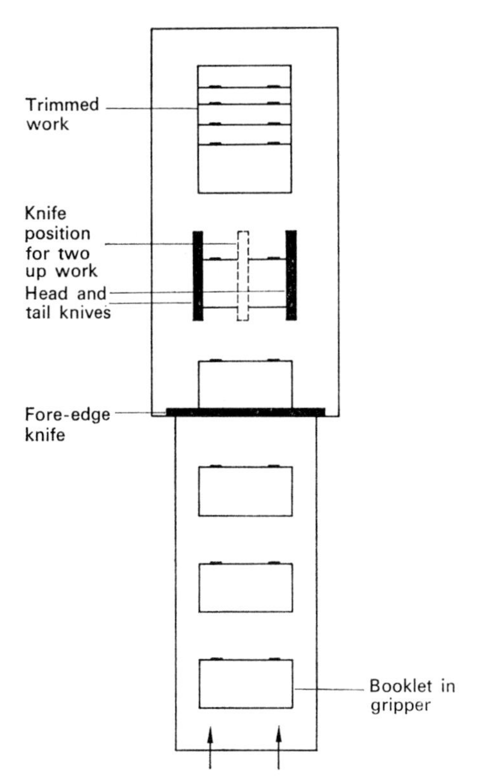

11.10 *Plan of single copy trimmer.*

12000–14000 copies per hour manufacturers have introduced a mechanism that doubles up the copies so that the trimmer need only operate at half the stitcher speed. This device is simply a rotary drum that picks up the first of two copies and deposits it upon the second book, the pair then proceeding into the machine which has all the features of a single-copy trimmer.

Automatic book feeding

Mention has already been made of the peg conveyors used in some multi-knife machines but standard pile trimmers are still offered with a hand-loaded in-feed clamp and for these a special automatic book feeder has been devised. This consists of a transporting conveyor which is linked through a gear set to the trimmer drive so that both units operate at identical speeds. When the book pile arrives at the trimmer a micro-switch activates a pneumatic ram that thrusts the book sideways into the in-feed clamp and up to the head stop. Air-operated joggers ensure that the pile is correctly registered before the clamp closes and feeds. Control of the ram pressure ensures correct feeding of all weights of paper stock.

Waste disposal

The removal and disposal of trimmings is an important part of the management of guillotines and trimmers. Sacks, trolleys and bins are used for small quantities but these are wasteful methods when many kilos are being produced weekly. The usual arrangements for single-knife guillotines and striker trimmers is a hopper located at the front of the machine by the work table and into which shavings may be put. In a simple installation this will be attached to a vent leading into a waste area immediately below; here the waste is sacked or baled for disposal. With a multi-machine installation, perhaps in several parts of the factory and at different levels, a more sophisticated approach is needed. Usually a vacuum system draws the shavings through a large-bore pipe system and deposits them directly into a baling press which may be electrically or manually operated. Most book trimmers have waste disposal arrangements to discharge directly into such a system (fig. 11.11).

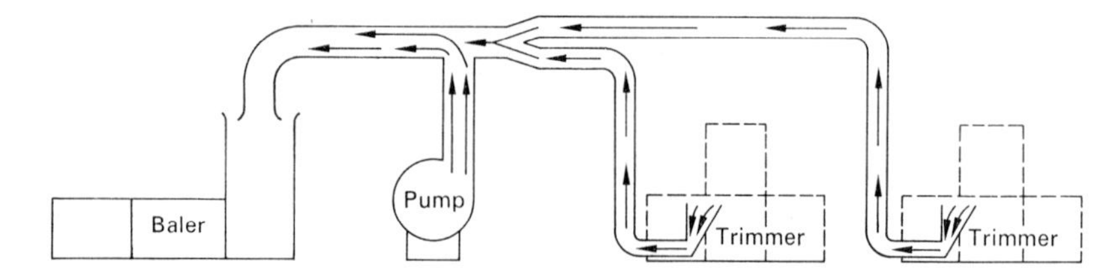

11.11 *Waste disposal system showing waste from two trimmers feeding one baling machine.*

Knife maintenance

The life of knives on guillotines and trimmers is directly controlled by the frequency of use and the nature of the stock being processed. In large centres the services of a trade grinder are usually available but some large users and those outside city centres may have, or prefer, to grind their own knives. A machine suited to this purpose has a horizontal bed to which the knife is clamped by screw or electromagnetic pressure (fig. 11.12). A grinding disc mounted in a traversing head is lowered on to the knife bevel at the required angle; the carriage then traverses along the knife, a standard knife length taking about 30 seconds, before automatically reversing. Coolant liquid is pumped on to the grinding area to prevent heating of the thin edge of the

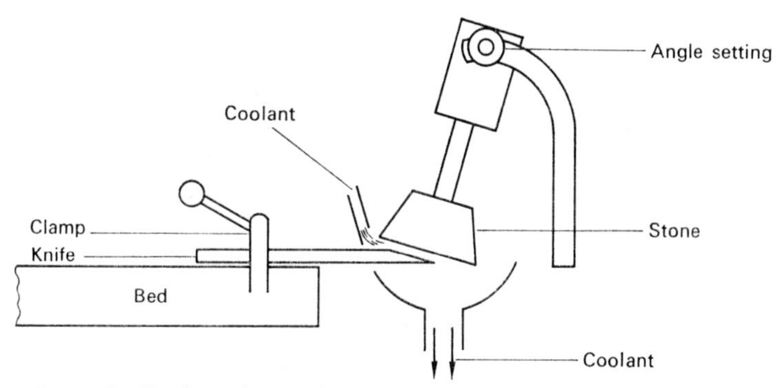

11.12 *Principle of knife grinding machine.*

knife. Some hand finishing may be necessary to remove burrs where these occur and it is important that minimum grinding is used.

MACHINE FOLDING

Folding machine running and production speeds have risen dramatically in recent years both in the jobbing and in the book and magazine fields. This has been made possible chiefly by better controls and improved engineering standards on the well-known buckle and knife folding principles.

On buckle folding machines considerable improvements of the plate have been made and adjustments are available to 1, open or close the throat by moving the surfaces of the plate apart, *eg* vertically; 2, alter their relationship horizontally with the folding rollers; and 3, tip the plate thus altering the angle of paper entry. Friction on the paper has been reduced by replacing the nearly solid plates with 'piano' wires or narrow aluminium bars. Accurate micrometer control is available at the front fold stop positions so that out-of-square paper may be easily accommodated.

Improvements on folding roller assemblies include better and more positive spring loading and control. Some rollers are set by simply placing the number of sheets of the stock being folded at that particular station between a calipering device and this automatically sets the rollers to the necessary pressure for that job. Experiments to improve the grip on paper without sheet marking include the provision of rollers with spiral milling, rollers with alternate bands of milled steel and inlaid rubber and others with a spiral rubber inset. The latter is claimed to both grip and smooth the sheet as it passes through.

To improve knife operation at high speeds the knives have a vertical action in long vibration-free guides and only a short movement. Sheet control mechanisms are more accurately machined and adjusted with sophisticated two-sheet detection on the feed board. Micro-switch safety stops are strategically placed and in this way expensive shut-downs are prevented. Machine speeds, governed by variable speed mechanism, up to 120m per minute are possible, and to utilise the full potential a device is incorporated to close the gap between sheets so that a nearly continuous stream of paper is passing into the machine.

Feeders may be of the continuous or pile types although the latter are most popular. These are suction drum feeders of improved design with rear separation by suction cup and blowers. Deliveries are usually of the stacker box or canvas belt types; in some instances the latter are independent of the main machine and may be wheeled to one of several positions as required by the job on machine. Very high speed production of book sections frequently requires special drum delivery mechanisms to clear the sections from the machine exit. Subsidiary attachments to modern machines include the usual perforators, slitters, creasers and gluing mechanism for 8- and 12-page work. Photo-electric counters and batch delivery devices are also available as extras. The improved engineering and design also includes needle bearings to fold rollers, central lubrication and variable speed mechanisms.

Advances in folding machine design discussed in this chapter refer chiefly to the so called 'continental' types and these call for considerably higher capital investment than do the standard varieties of the same size. These machines are principally in the small and medium size range and their use is a natural corollary to the trend in small high-speed printing machines. Where very large sheets are printed, *eg* in book production, it is necessary to split the sheets (on the printing machine or guillotine) before folding on the continental folder. However, books are often printed four or more sections on one sheet and as quad folders are available to handle the large sizes they continue to be used

One application of the all-buckle folder in the quad folding field takes paper printed half-sheet work and slits it into half as it enters the first fold assembly. The lay half of the sheet goes up the first fold plate whilst the off half goes down the second fold plate. Each half of the plates not being used has a deflector fitted. This, of course, delivers the two units to the first cross carrier, almost simultaneously, for the next right-angle fold, but as this is also a plate fold no timing cycle is necessary; with appropriate imposition 2 or 4 × 16 page sections are obtained.

Quad folding systems

The term 'quad' folding implies folding large quadruple size sheets of paper, usually into book or magazine sections. Standard machine sizes may be quoted from 1016 × 762mm and up to 1320 × 1930mm, but as these are maximum sizes some machines will fold sheets as small as 381 × 254mm.

Most all-knife machines in this range are fed from a continuous feeder with combing or suction wheel separation, the former being preferred. Delivery to the folder is via drop rollers working in timed cycle with the knives; up to 3500 per hour running speed is achieved. Each folding station is similar to the others and consists of transporter tapes, front-lay stops, side pull (or push) mechanism, vertically acting knife, two fold rollers and associated fold bars. The sheet control method used at each level may vary to suit thickness of stock and number of folds already incorporated.

Perforator knives are mounted in a fixed position in the folding roller on all right-angle layouts; the fold actually occurring along the perforated line at the next fold station. One type that folds the C and C* imposition has a bar perforator sited immediately before the second of two parallel folds. Without this the head of the work would not be perforated and creasing would result. Slitters in single or ganged assemblies are normally sited at the point where the folded sheet is ejected into the delivery box; there are, however, some machines that slit the sheet between folds *eg* imposition G.

All knive machines are often classified as either jobbing quad folders or as book folders; the former may be able to fold 8-, 16-, 24-, 32-, 2 × 16- and 2 × 32-page sections in a wide variety of sizes, while the latter may be confined to one or perhaps two alternatives only in a limited size range. Jobbing folders often have several delivery positions around the machine and the sheets are directed into the required sequence of folds by strategically placed plates. An alternative method to redirecting is that of 'sheet turning'; this device grips the sheet and turns it through 180° so that the following knife may be used for both cross and parallel folds (fig. 11.13).

One series of machines can be built up from a basic jobbing machine with capacity previously mentioned into a complex folding system capable of producing a wide range of sections. This is achieved by the addition of extra folding units, at any time after installation, which, for instance, may add 4 × 16- and 8 × 16-page schemes to its potential.

Most book folders are equipped with needle position in the first and second fold knives; these are usually retracted for normal work but if a job printed on thin paper tends to skid on the impact of the knife, the needles are lowered so as to skewer the paper just before the knife makes contact. This ensures that the paper is quite stationary before the knife drives it between the rollers.

Another useful device in certain circumstances is one that will add the operation of gathering to the folding machine. This is applicable to 128-page

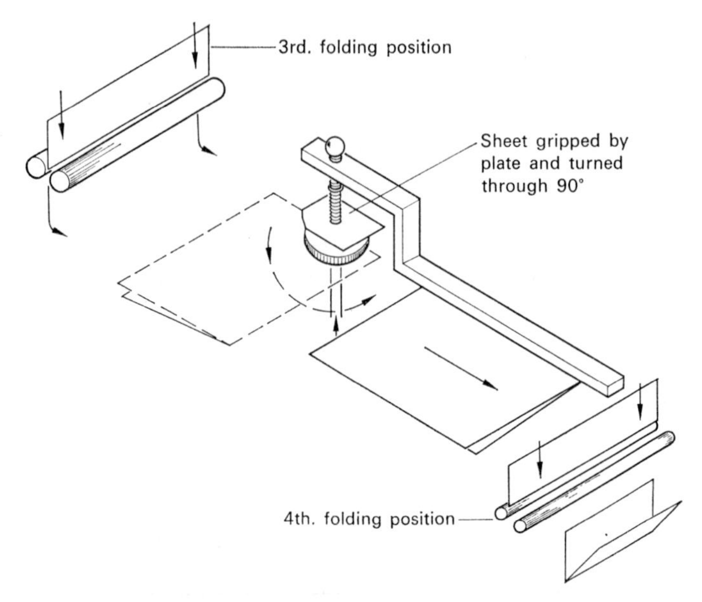

11.13 *Sheet turning mechanism of knife folder.*

sheets printed 8 × 16 pages up, parallel folded, and gang slit to deliver eight consecutively numbered sections (scheme X). These are delivered in staggered formation and an arm sweep gathers them into a stacker box as a complete book. A 256-page book can be gathered in 32 pages in a similar way, but of course with a maximum sheet size of 1320 × 1930 mm the book size cannot exceed 165 × 121 mm (fig. 11.14).

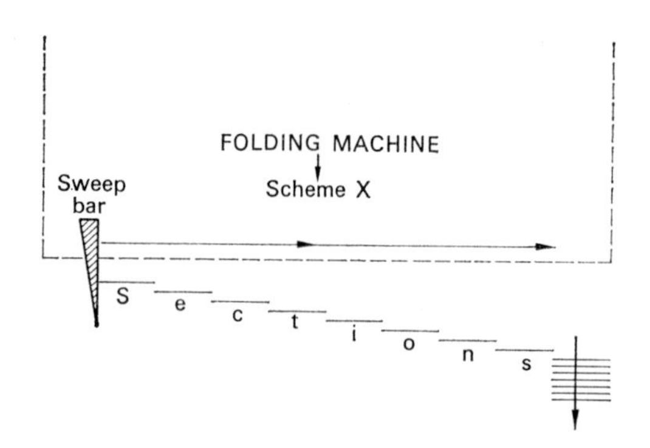

11.14 *Gathering a 128-page book on a folding machine.*

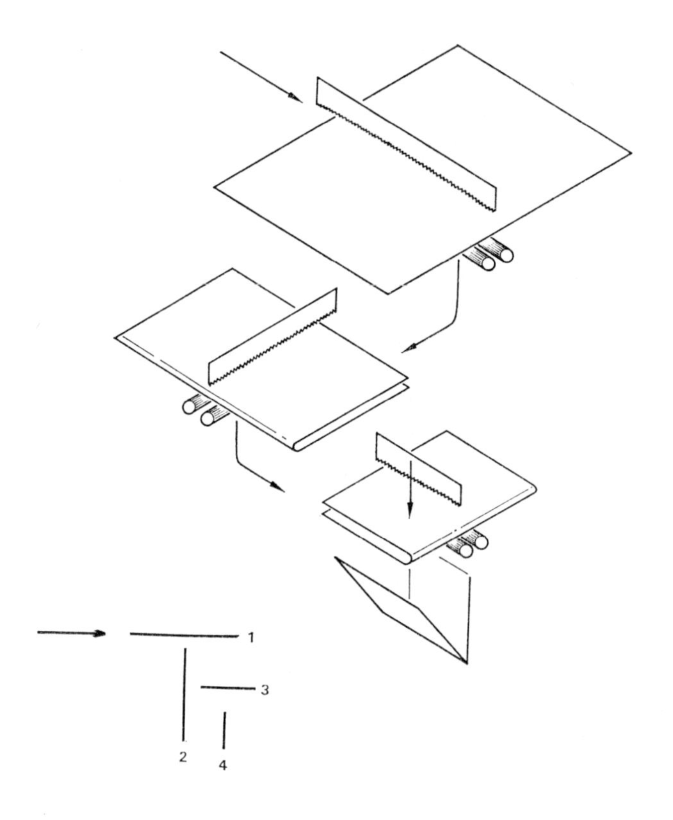

11.15 *1 × 16 pp; layout and folding of scheme Z.*

On the 'duplex' machines two continuous feeders are mounted in tandem so that, for each cycle of the folding machine two quad sheets are fed and registered under the first fold knife. Applied to the C and C* type of machine it makes possible the delivery of four 32-page sections with bolted heads.

Modern all-knife machines have variable speed drives, central pressure lubricating systems, electrical inching control buttons, self-tensioning tapes and counting and stacking devices. Problems on quad folders may be summarised under three heads: grain and caliper of paper, machine control and printing quality.

Criteria used to determine whether, on a specific job, long or short grain bookstock should be used seldom includes those factors that influence the folding problem. If this runs parallel with the direction of the movement of the sheet then a rigid leading edge is presented to the front lays and this may result in 'bounce'; this is particularly relevant to thick and stiff types of paper. Very thin and flexible papers tend to crumple when striking the front lays and bad folding may result from the random way this occurs. Accurate setting of dabbers or slow down wheels, sheet control brushes and regulation of the machine to an optimum speed for that particular stock will all improve the chances of successful folding. After the first fold has been accomplished control over the paper becomes much easier. Movement of the paper from fold to fold is partially governed by the pressure on the folding rollers and these must be accurately set to give correct position of the perforator cut as this controls the squareness of the subsequent folds. Accurate setting of the delivery rollers is also essential when delivering and slitting a multi-section imposition or the heads of the sections will be out of square.

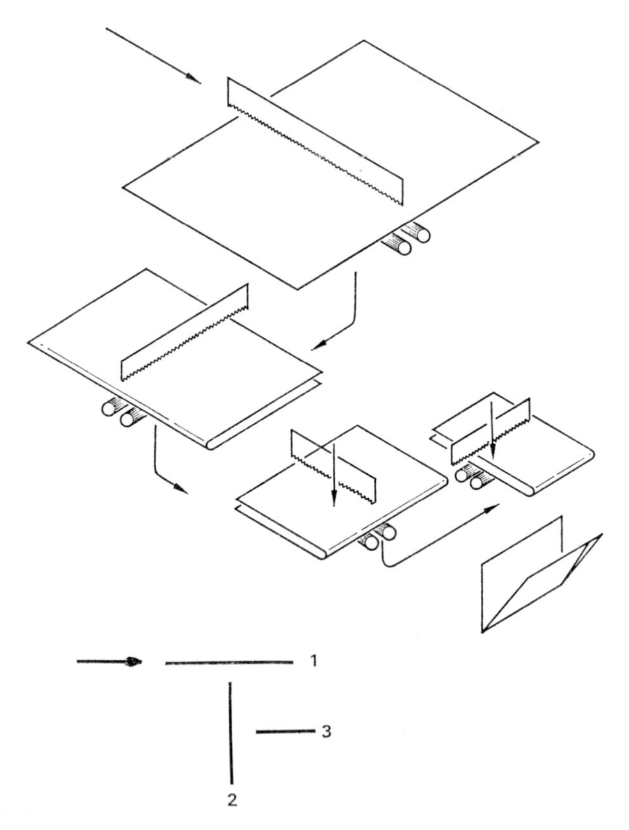

11.16 *1 × 32 pp; layout and folding of scheme M.*

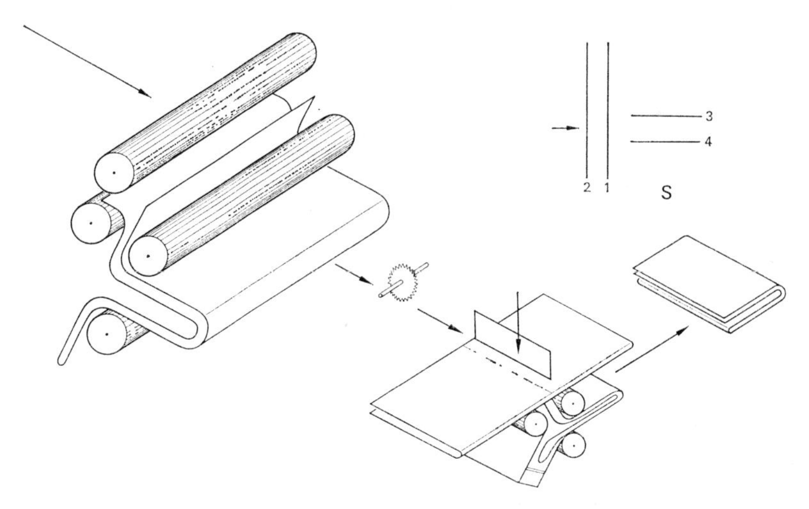

11.17.1 *1 × 32 pp; layout and folding of scheme S.*

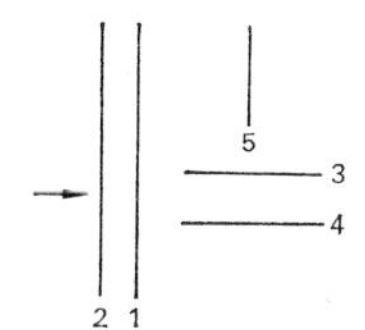

11.17.2 *1 × 64 pp; layout for scheme T.*

The printing of a tack mark on the sheet edge considerably enhances the folding machine operator's ability of assessing the folding situation before the problems arise. Close examination of tack mark position throughout the stack will readily show the accuracy of printing grip and lay edges. When work has been printed two up and split on machine the accuracy of the split may also be displayed by the tack mark. Whilst minor variations may be ignored, bad discrepancies may be avoided by sorting the sheets before loading on the feeder.

As most book work is printed on mill-slit paper variations in size are not uncommon. Although by the nature of things the variation will be on the leave edge this may effect the quad folder when critical control brushes or rollers are set exactly on the tail of the sheet.

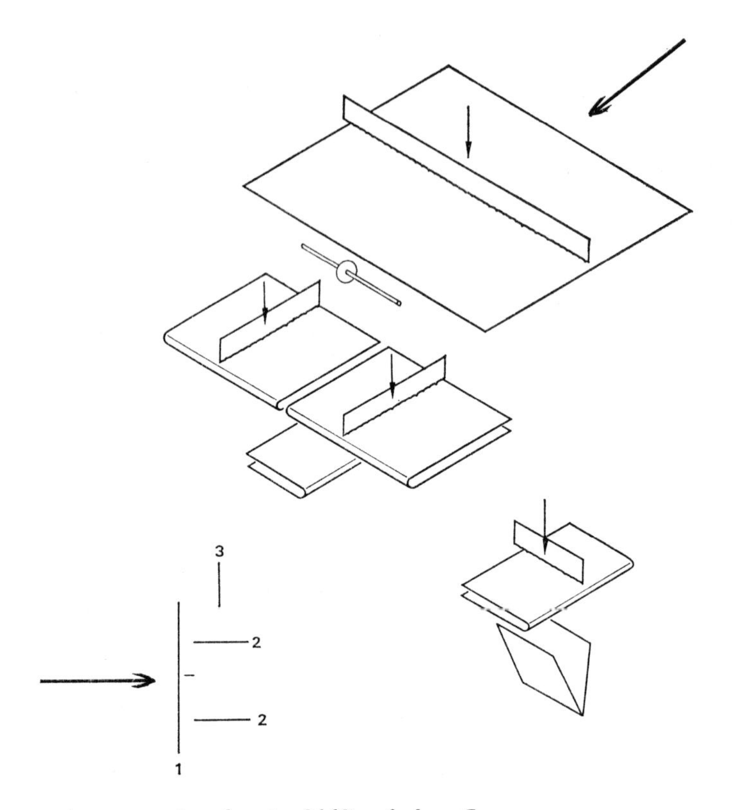

11.18.1 *2 × 16 pp; layout and folding of scheme G.*

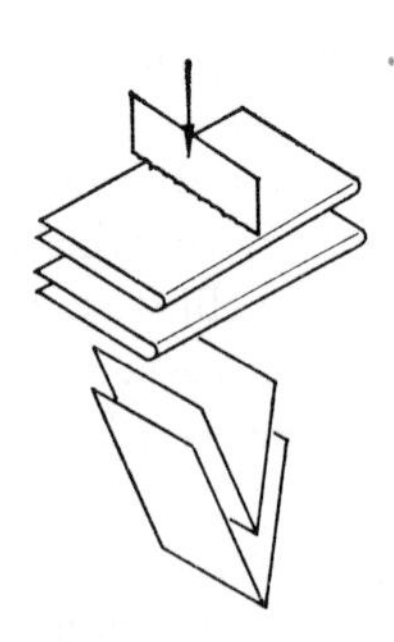

11.18.2 *Both 16-page sections are assembled under the third knife to deliver a 32 pp (16 inset 16) for scheme F.*

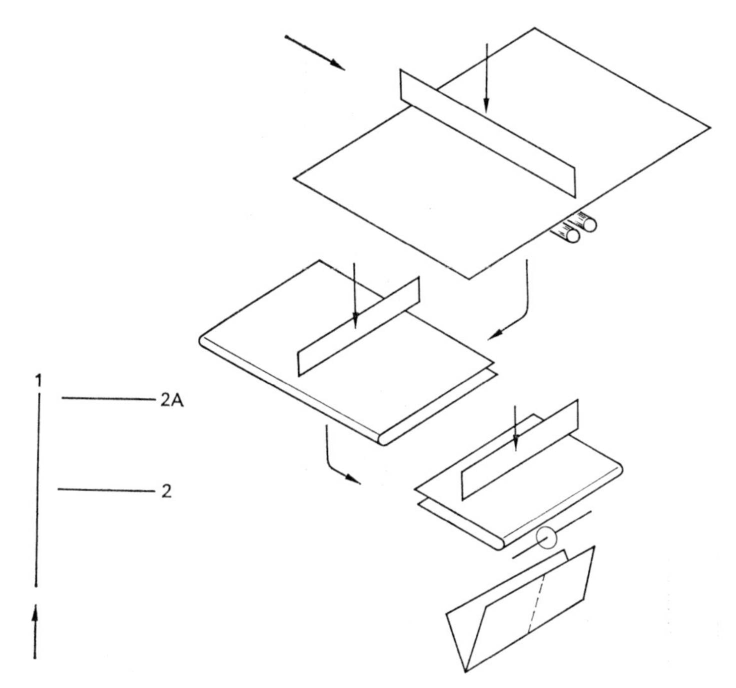

11.19.1 *16 pp, 2-up; layout and folding of scheme U.*

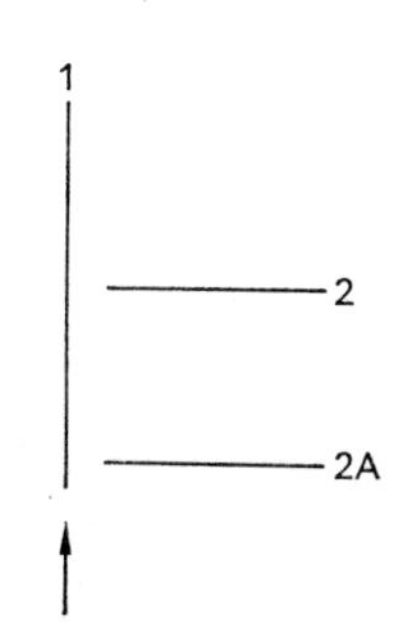

11.19.2 *16 pp, 2-up; layout of scheme W. (Note alternative direction after second fold.)*

BFMP BOOK IMPOSITIONS

Book and magazine sections printed on conventional sheet fed machines may be imposed to have any one of a number of varying impositions to suit the job in hand and the folding and other processes that follow. Most sections contain either 16 or 32 pages, but under certain circumstances 12 and 24 pages

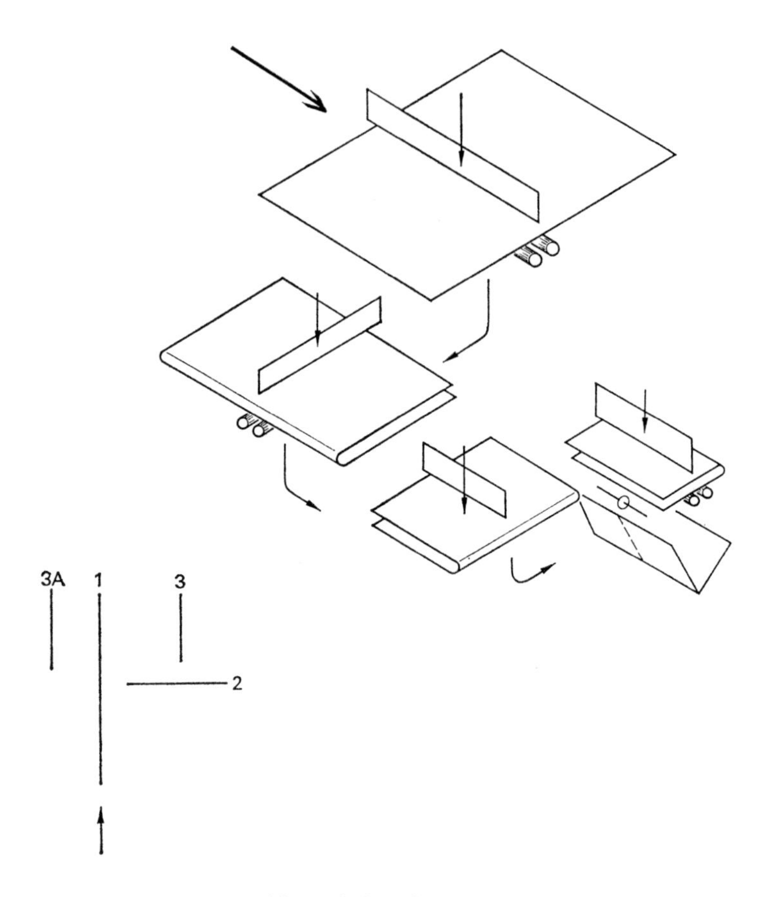

11.20 *32 pp, 2-up; layout and folding of scheme O.*

may be imposed and folded. The latter may be particularly useful when a job is of such a format that 16-page sheets would waste printing and folding machine area whilst 32 pages is too large for the machine.

The British Federation of Master Printers has classified all the major book and magazine impositions allotting a letter of the alphabet to each one for ease of reference. This is very useful when the job is to be printed in one building and bound in another. The most used folding schemes as they appear in the finishing department are shown on page 164. The letters of the alphabet that appear in the manual, but are not in the summary, are those alternative methods of imposition that usually require slitting on the printing machine or cutting on the guillotine before folding.

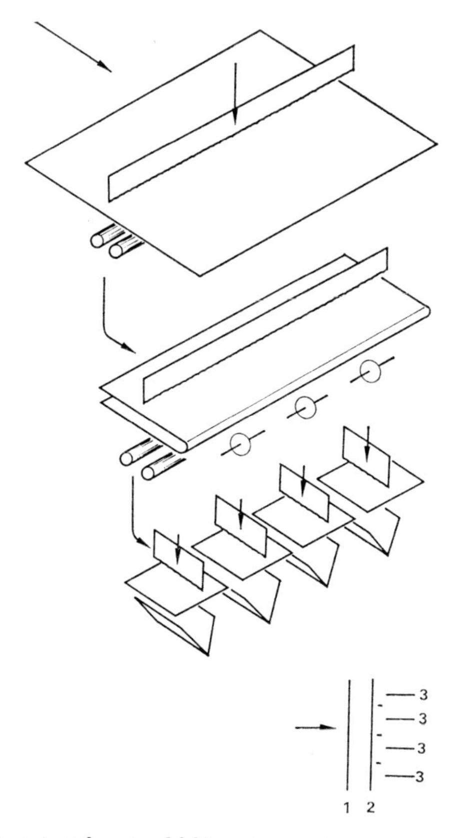

11.21.1 *4 × 16 pp; layout and folding of scheme C.*

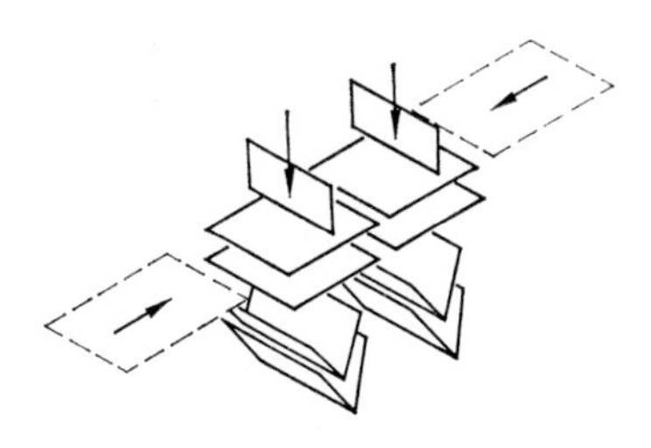

11.21.2 *For scheme C* the outside 16-page sections move inwards so that the two centre last fold knives can fold as 2 × 32 pages (16 inset 16).*

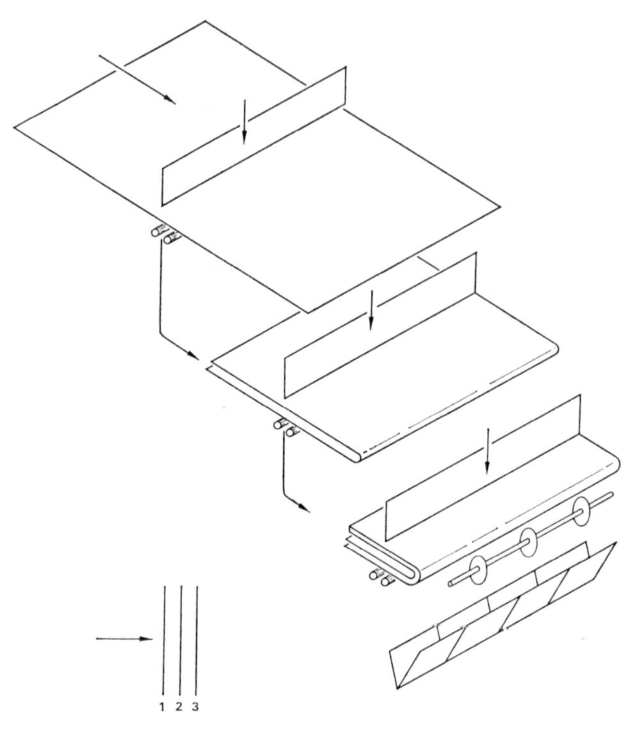

11.22 *4 × 16 pp; layout and folding of scheme P.*

Reference to the manual shows that schemes Z (fig. 11.15) and M (fig. 11.16) are conventional 16- and 32-page right-angle folds respectively. The alternative scheme S (fig. 11.17.1) is one that is found on combination and 'continental' type folders and produces a single 32-page section having a double-bolted head and bolted fore-edge. Scheme T (fig. 11.17.2) produces a single 64-page section and is unusual in standard bookwork, as it has limitations of size and paper caliper in application.

Table 2 Summary of BFMP folding schemes

Output	*Scheme*	*Output*	*Scheme*
1 × 16pp	Z	1 × 32pp	F, M, S
2 × 16pp	G, U, W	2 × 32pp	O, C*
4 × 16pp	C, P	4 × 32pp	Y
8 × 16pp	X	1 × 64pp	T

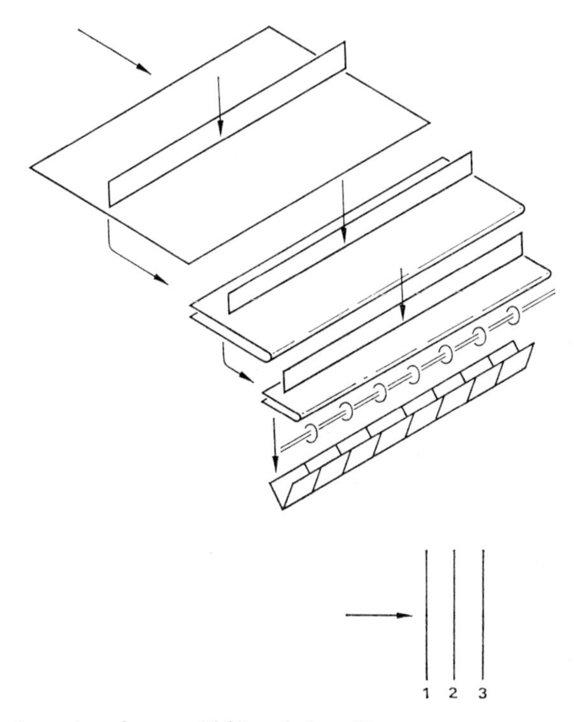

11.23 *8 × 16 pp; layout and folding of scheme X.*

Simultaneous folding of two 16-page sections is obtained by using schemes G, U or W. The first of these (fig. 11.18.1) slits the sheet into separate units between the first and second folds while the other schemes complete the folding before the split is attempted. This means that in the latter method there is a choice of delivering two separate 16-page sections or one unit of two sections unsplit. Leaving the work two-up is often a good way of producing saddle stitched, wrappered and, where size permits, sewn case bound work. Similarly the folding of two 32-page sections is covered by scheme O, the same general provisions applying (fig. 11.20).

A single 32-page section is produced by scheme F when the double sixteen machine (scheme G) is adjusted to inset sixteen into sixteen on the machine (fig. 11.18.2).

Four 16-page sections are obtained by using schemes C (fig. 11.21.1) and P (fig. 11.22). Scheme C is deservedly popular because it delivers sections with bolted heads, and to achieve this the sheets are split between folds 2 and 3. This necessitates having four last fold positions or having two knives that

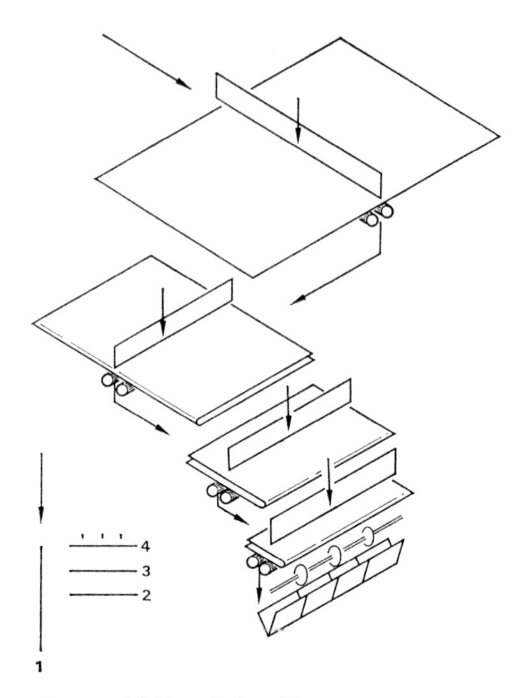

11.24 *4 × 32 pp; layout and folding of scheme Y.*

make a double movement. Scheme P produces four 16-page sections from three parallel folds, the folded sections being split into four units just before delivery. Eight 16-page sections are produced from a similar layout of folding rolls and a sheet containing 128 pages by scheme X (fig. 11.23). Again the slitter reduced the paper to a single or two-up unit, as required between last fold and delivery and gives sections with open heads.

By suitable arrangement of the machine scheme C may be adjusted to allow the 16-page sections to inset and deliver as two 32-page sections, the outside 16-page moving inward in each case. The resulting scheme C* section has a double bolt at the head (fig. 11.21.2). Four 32-page sections produced by scheme Y are slit at point of delivery and have open heads (fig. 11.24).

Factors that govern the selection of suitable folding schemes for a particular job include the size of the printing machine and its relation to the size of the finished section. A sheet of A2 paper will produce one 16-page section of A5 size, folding scheme Z. If the printing machine can accommodate A1 paper, then two alternative methods of folding become available: 1, print 2 × 16 pages (scheme H or HI) to split on machine or guillotine and fold as scheme Z; and 2, print and fold 2 × 16 pages imposed U, UI, W or WI.

Another consideration is that the hand feeding of sewing machines and insetter-stitchers requires that the sections have easy access to the centre position and the most successful folding schemes are those having bolted heads. The 'rate' for hand-feeding open heads is usually higher than for bolted heads and this reflects the extra difficulty in feeding this class of work. Where possible the provision of projecting bolts on the fore-edge will greatly assist the feeding operation but this must be anticipated at the imposition stage.

Tipping-in 2-page plates is sometimes required at points other than the natural opening places of sections and folds may have to be split by hand to allow access. Some folding schemes, having open heads or reduced numbers of fore-edge bolts, may be more economic in this direction than others.

Although for diagrammatic purposes the schemes are illustrated as folded on an all-knife machine, it must be noted that many of the schemes are suitable for buckle and combination machines.

IN-LINE PRODUCTION OF SADDLEBACK MAGAZINES

For the high-speed production of magazines and booklets fast insetter-stitcher-trimmer combinations are used. These may be hand loaded or fitted with mechanical feeders (see Chapter 10). Sections are laid on to saddle bars or 'swords' and as the conveyor pin passes underneath, the section is lifted off and transported towards the stitcher head. An operative or automatic feed-head is needed for every section or unit in the make-up and this clearly has a strong influence on the economics of production, make-up always being in the smallest number of units possible.

Transfer from the chain to the stitcher is effected by grippers and a trip operates to prevent the stitcher heads functioning if no copy is present. Automatic machines usually have a caliper device which monitors the thickness of the books passing through; ejection occurs if the book is too thick or too thin. Stitching heads may number from two to six depending upon the work normally completed and in some instances a reciprocating action is given to the heads ensuring a smooth flow of work from the chain to the delivery box. All suitable standards of wire are used and some machines will stitch up to 6mm of paper, *ie* a book 12mm thick (fig. 11.25).

Delivery is into a packer box for off-loading by hand or directly to the single-copy trimmer (see this chapter). As with other in-line methods high speed insetter-stitcher-trimmer combinations are frequently linked into a complete production system that may include inserter, batch counter and packing

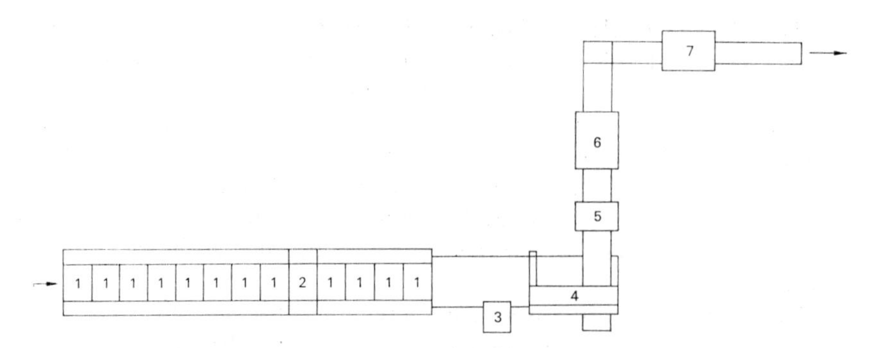

11.25 *Plan layout of insetter-stitcher-trimmer-inserter combination.*

1 *Individual feeders*
2 *Subscription feeder*
3 *Book caliper safety device*
4 *Stitcher*
5 *Book collecting device*
6 *Multi-knife trimmer*
7 *Inserter*

unit. Other on-machine attachments that are found on sophisticated equipment include a subscription-card feeder and centre-spread paster.

The inserting machine may be in-line or can be operated as a separate hand-loaded unit after the book (square or saddleback) has been trimmed. It is equipped with one or more rotary feed heads mounted above a conveyor chain. As the book moves along the conveyor the uppermost cover is opened by suction pads, and, as it passes the feed head, an insert is placed in position; by inverting the book in the feeder either end can be opened. It is also possible to open the book with a knife blade and to place the insertion, but not at a predetermined page. When used as a separate installation this machine also feeds the books that are to have the insertion.

A very wide range of insetter-stitcher combinations are offered, from small two-station versions running at a maximum of 4000 copies per hour to the automatic lines having a number of automatic feed heads with maximum speeds of up to 12000 copies per hour.

12. Publishers' binding

The term 'publishers' binding' or 'edition binding' is used to indicate that a quantity of books is to be bound in the case style for a publisher who bears the cost of printing, binding and other expenses in anticipation of a profit deriving from the sale of the volumes. Length of run of such work varies from a few hundreds to hundreds of thousands and clearly the problems involved in processing such different quantities may be extensive, as it is not usually worth while setting up automatic machinery for very short runs.

The main concern of this chapter is the handling of average length of runs of work that is of average size, but it must be understood that all productive machinery has upper and lower limits for length, breadth and thickness of books it will process. Work falling outside these limits may need to be the subject of special machinery providing the quantities are sufficiently large, *eg* diaries. Occasional runs of very large or very small work will be partially hand assembled.

The organisation of publishers' work is usually arranged around the 'dummy' book. This is a copy of the job bound in the factory some time beforehand to prove the materials and sizes. Usually two or more cases are blocked, one being retained as the copy for the actual run. This may give the publisher the opportunity to have second thoughts on choice of cover, endpapers, blocking, etc.

'Early' copies are often requested by the publisher and are used for promotional purposes and reviews, etc. The difficulty of producing a short run like this can be appreciated and sometimes special rates are charged for this service.

During the planning of production certain facts about the job will need to be known and a typical order sheet is illustrated (fig. 12.1). From this sheet departmental instructions are made up and distributed.

Folding

Folding may be divided into three groups; text, plates (illustrations) and endpapers. The BFMP imposition system forms the basis for text folding and

WORKING INSTRUCTIONS

CUSTOMER A. B. Publishing Co. ORDER NO. 46208

PATTERN BOOK NO. 3/69 PRINTER YZ Printers

TITLE Philosophy in the New World SIZE A5

No. of sheets in	5000	Bind 1000
Imposition (scheme)		C*
Make-up		1 x 16, 10 x 32
Overcast first and last section		No
Endpapers		Cartridge
Ends reinforced/plates		no — nil
Sewing		French
Bulk		20 mm
Trimmed text size		A 5
Weight of board		1500 gm^2
Cloth ordered/reserved	yds	supplied pre-printed
Brasses (who supplies)		Publisher
Blocking: Ink, gold, foil		Alumin. foil
Position		Title + Author
No. of runs		1
Calico lining		no
Edges		plain
Round and back or square		R + B
Headbands or marker		no
French or tight grooves		French
Jacket		Yes
Inserts		1
Packed		dozens
Where delivered		Pub. warehouse
Contract date		No
Remarks		

12.1 *Working instructions sheet.*

quad folders folding C or C* will process between 60 000 and 100 000 quad sheets per week; it is therefore not at all unusual for one machine to produce one third of a million 16-page sections in a forty-hour working week. Although similar running speeds are used on other folding schemes the section output may be lower because only one or two sections are being produced at each cycle of the machine.

Plates, endpapers and odd sections are folded on all-plate or combination folding machines. Where applicable the two-up method is applied, slitting into singles on delivery. This is particularly relevant to the folding of endpapers for books in the smaller octavo sizes.

Bundling

This is a relatively simple operation made necessary by the large quantities of sections that have to be stored until gathering can commence. Folding a 10 000 job of 320 octavo pages imposed in 32-page sections will require five quad sheets, *ie* 50 000 sheets in all. Using only one machine this job will take about three days to fold and it will not be until the second half of the third day that gathering can commence. Meanwhile the product of the first two days' folding have to be accommodated and the sections are bundled and stacked.

The machine is a trough-shaped press having one fixed and one movable end. The folded sections are assembled in the trough with thick boards at each end and the pressing head actuated. Whilst under pressure a loop of string is passed around the bundle and tied off. When the machine pressure is released the string retains the bundle of sections under pressure.

Whilst the bundling is primarily to enable convenient handling, the compression does of course compact the sections squeezing out air and consolidating folds. Although the general effect of this pressing is good there can be some bad side effects. Most tying of bundles is effected around the long edge and there is really very little control around the head and tail parts of the sections, which can be distorted during the handling of the bundles. This distortion is a major factor in the operation of gathering machines as it will prevent the sections lying flat in the hopper and cause malfunctioning of the selector mechanism. To obtain best results in bundling heavy boards and wide tapes may be used and this will ensure good undistorted sections. Another alternative that may be used on machines that will not take wide tapes is to tie the bundles around the short edge of the sections. Some companies have made attempts to dispense with bundling altogether but in binderies where perhaps more than a million sections are folded daily this becomes very difficult to organise.

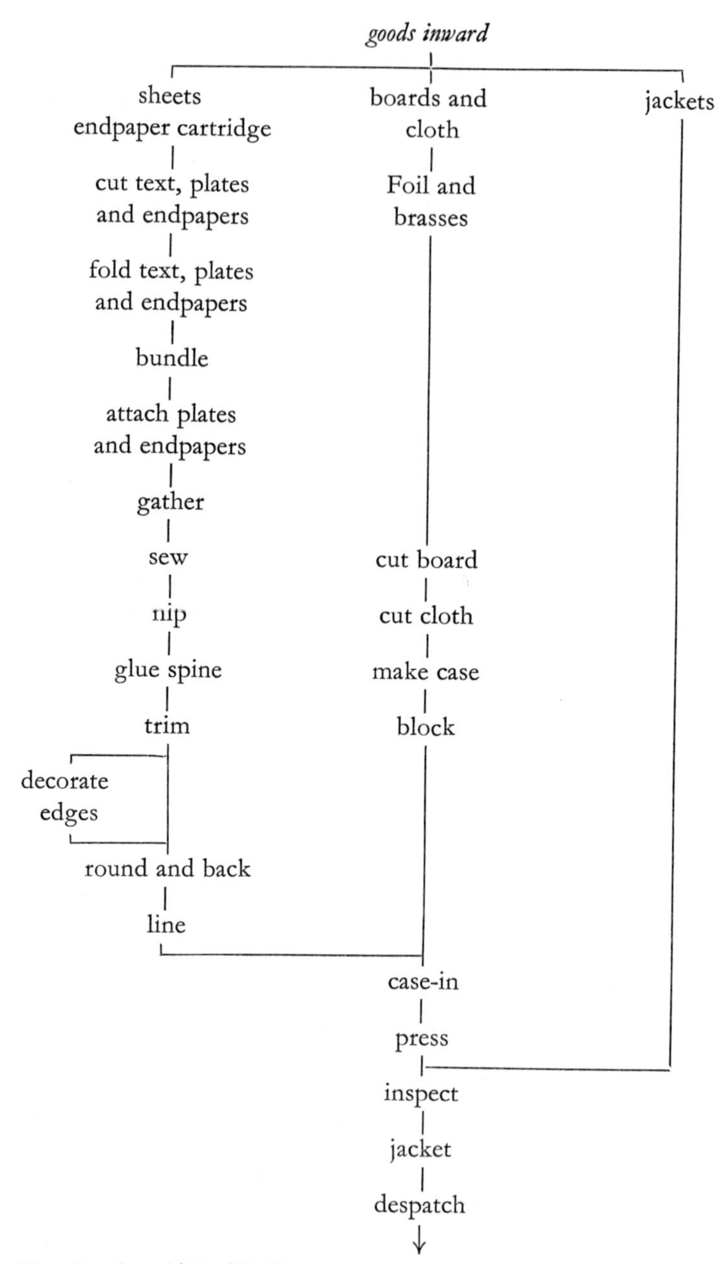

12.2 *Flow chart for publishers' binding.*

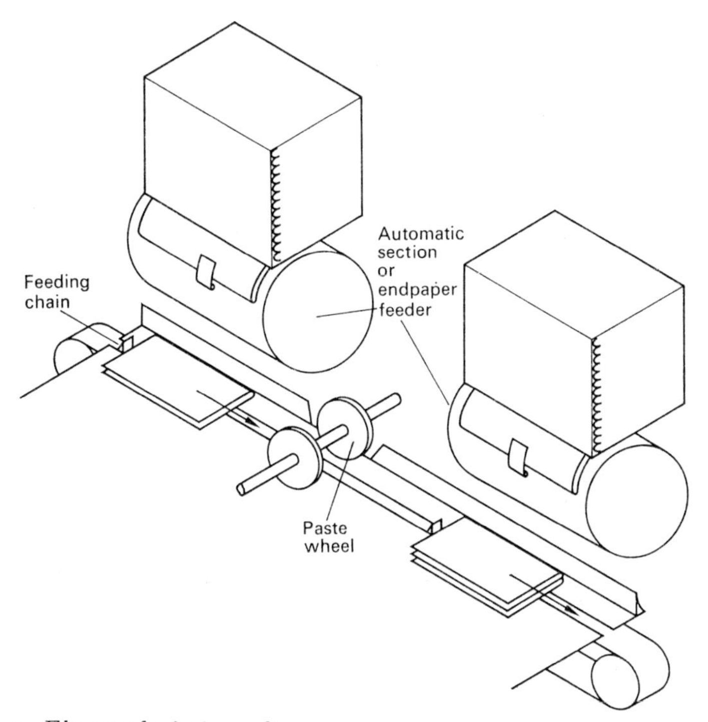

12.3 *Elements of a tipping machine.*

Bundling presses may be hand, electrically or pneumatically operated, the power machines having adjustable pressures. Bundles are labelled to show their contents and stacked on pallets sorted ready for either direct movement to the gathering process or to the plate and endpaper attaching area.

Attaching plates and endpapers

Endpapers for this style are simple tipped-on four pages. Machinery for this task may be hand or automatic fed, the latter predominating. The endpapers and the sections are loaded into adjoining hoppers and the separating mechanism selects the bottom unit from each and delivers it into the trough. A chain carries the endpaper along the trough and past a pasting unit which deposits an edging of adhesive about 3mm wide on the fold edge. When the endpaper reaches the second feed position the section is neatly positioned on top and the assembly passed between pressing rollers before delivery. The pasting mechanism may be either the wheel or bar type and the adhesive is starch, gum or PVA formulations of stiff consistency (fig. 12.3).

When the book is to be adhesive bound it is necessary to tip the endpaper away from the spine fold the depth of the cut to be made. In this way the completed book shows a fold in the endpaper opening instead of the glue line that would otherwise be the case.

Attaching illustrations, *eg* 2- or 4-page plates, may also be effected on these machines; the relatively slow hand-fed varieties are often used but the automatics do tend to have accidents which result in the spoilage of the printed illustrations. As the percentage of overs is strictly limited, the binder can seldom risk attaching plates by machine because of the possibility of coming up short in totals. Today, the trend in illustrating bookwork is away from tipping and toward insets and wraprounds.

Gathering

The thinnest gathered book will be of three sections whilst the thickest may be 60, 70 or in extreme cases 100 sections. Gathering machines may be divided into two basic types usually referred to as 'arm' and 'rotary' and this refers to the manner in which the sections are transferred from the hopper to the conveyor; but some machines do not easily fall into these categories.

In all cases the sections are placed in the hoppers in a continuous row, each hopper containing from 100 to 300 sections according to thickness. The section at the bottom of the pile is drawn clear of the stack by a set of sucker pads, gripped and withdrawn from its position. It is then deposited in a raceaway so that a chain or conveyor can collect it. As the last section is deposited on the belt first and the first section deposited last, a complete book is delivered when the conveyor has traversed the machine.

The sections within a book are often of different thicknesses so it is necessary for each feed position to be set individually for caliper. When correctly set the gripper will metre the section and only continue to function if the caliper is correct to the setting. Should a double feed, misfeed, thin or thick section appear the machine will come to a halt. When stopped an indicating lever or lamp shows where the stoppage has occurred. Theoretically the machine is foolproof and cannot misgather but in practice the hoppers are sometimes loaded with the wrong sections and this necessitates a percentage check on the product of the machine. If one incorrect book is found it is likely that there will be others in front and behind the faulty book.

Machines are usually available in units containing, say, three hoppers and machines can be made up in length to suit the customer's requirements. A suitable number of hoppers for a book house is 20–25. Books of 6 sections may be gathered 3 or 4 up and books of, say, 35 sections would be gathered in halves and married up after the second run.

Sewing

The majority of publishers' case books are sewn using the french or plain sewing method and although not as strong as tape sewing has the advantage of high production speeds and relatively low cost. Some publishers insist upon 'over tape' or 'through tape' for their best work and are prepared to pay the quite high difference in the cost for the extra strength obtained. The higher charge is brought about by slower sewing-machine speeds, the need to paste the tapes to the endpapers and the extra difficulties encountered during forwarding.

Three types of machine are in common use and these differ mainly in the method of feeding the sections. 'Hand fed' implies that the sections are fed directly on to the saddle and up to a gauge by the operator. Because of the short time at bottom dead centre the operating speed needs to be relatively low to give the operator time to locate the section in position before the saddle rises. Even so, on bolted heads and fore-edges, production speeds in excess of 2 000 sections per hour are usual. Many of the older machines still require that the books be cut off by hand and pasting of first and last sections is also effected by hand. In most companies all tape sewing is carried out on this type of machine.

A sewing machine that relieves the operator from placing the sections directly on the saddle is referred to as semi-automatic. A feeding station sited at the side of the machine is loaded with the section and a transfer mechanism delivers it to the machine saddle when this is at the bottom of its stroke. As the machine takes care of the heading of the section the operative is free to concentrate on the feeding and improvements in production of more than fifty per cent over the hand fed types are obtained. The cutting-off mechanism contains a sharp pointed knife edge which engages the blind stitch that occurs between books, allowing them to be pulled apart after delivery. A foot-controlled bar pasting mechanism is utilised at will to strengthen the first and last sections. This is the most widely used type of machine for sewing publishers' case bindings; some of the models currently offered provide facilities for french sewing only.

Automatic sewing machines have been available for some time but their success has been limited by their inability to open the wide variety of section make-up that may be found in any one book. A long run of an unsophisticated book having self-endpapers and no other plates could be imposed specifically for the auto sewer and a successful run accomplished. Another job having tipped endpapers, plates insets or wraprounds and perhaps odd sections, *eg* 8 or 12pp, would require so much organisational work to prepare it for the sewer that the savings made by any increase in production would be lost in the earlier preparation work.

Problems of section feeders are dealt with in a previous chapter, but a new and rather novel type of feeder has recently appeared on the market. This embodies a memory belt that controls the operation of the machine. The links of the belt are re-assembled to suit each different book make-up, each link relaying instructions to the machine in relation to the method of opening that should be used for that particular section, *ie* fingers or suckers.

Thread used on sewing machines is made from cotton, terylene, nylon or a blend of cotton and terylene. Cotton works well on sewing machines but for publishers' work a caliper of around 0·25mm is needed to obtain a suitable breaking strain. This tends to be bulky and causes swell problems in books of many thin sections. Both terylene and nylon threads have been used and are very strong even in thin calipers; however, because of the elastic nature of these threads tight sewing is difficult to accomplish. The blended thread provides a material having a good tensile strength for its caliper of 0·125mm and the workability of cotton.

Threads are available in various white and unbleached shades wound on to conical cops of different weight sizes. As threads can be spun with the right-hand or a left-hand twist it is usually necessary to follow the manufacturer's instructions on the correct selection of threads.

If publishers' books are sewn upon tapes these should be of the thin open weave variety so that the paste used in the casing-in process may penetrate and secure them to the endpaper. An alternative procedure is to paste the tapes into position, immediately after sewing, leaving some slack to be taken up by the backing process.

Nipping

This process consolidates the books and brings them to uniform thickness, but excessive pressure should be avoided if the volume is to have a good shape. It is usual to nip the work to fit a case previously made to correct size.

Nipping or smashing machines are powerful presses having a continuous reciprocating motion, the work being entered or removed whilst the jaws are apart. The smallest machines have a horizontal or angled action with the jaws only 50–100mm wide. This means that only the spine portion of the book is compressed and this is often inadequate for volumes with bulky fore-edge folds. The alternative vertical acting presses have large area jaws and will provide the all-over nip needed for difficult work. Although these machines have relatively low operating speeds the work is processed in piles of 100–130mm high. Thus a book bulking 30mm would be nipped in groups of four books (fig. 12.4). Attempts have been made recently to integrate nipping into an automatic line by treating each book singly.

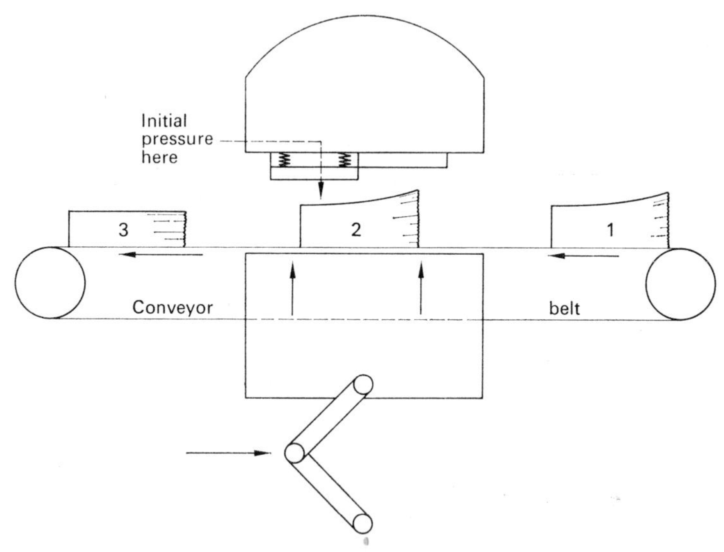

12.4 *Automatic toggle-operated smashing machine having three stations* (1) *feed* (2) *press* (3) *delivery on a leather conveyor belt.*

Spine gluing

This operation may be completed either before or after trimming the edges; the current trend is toward the former because of the introduction of automatic gluer-driers with attendant difficulty of maintaining smooth edges on trimmed work.

Small hand-fed machines apply the adhesive by roller, brushing it between the sections and removing the surplus. It is important that the viscosity be adjusted to ensure penetration; very thin sections require a thin adhesive whilst the same adhesive used on thicker sections would allow it to penetrate and spoil the work. Both PVA and hot animal glues are used; in both cases the adhesive should be of top quality so that lowering the viscosity by the addition of water will not seriously affect its properties. Another means of lowering the viscosity of hot animal glue is to raise the temperature and this may be a better method than adding water. The cohesive and flexing qualities must be good too; when subjected to the rounding and backing operation tremendous strain is placed upon the ability of the adhesive to maintain the sections in a single unit. After gluing in groups the books are 'turned', *ie* stacked with spines in alternative directions, to dry. This may take two hours or more before the work is ready for trimming.

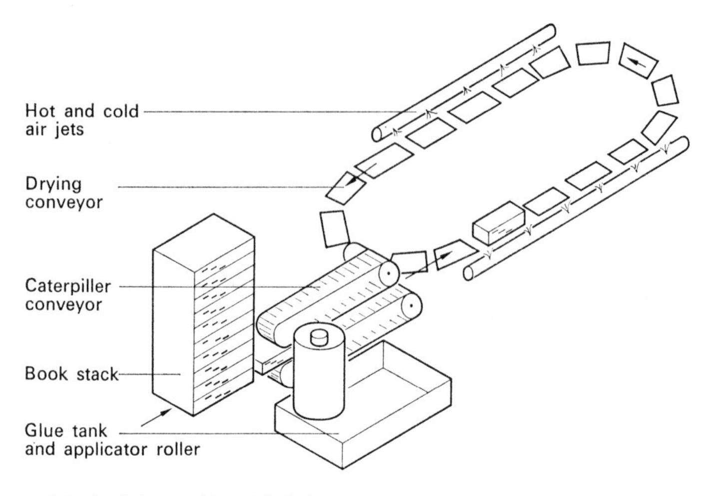

12.5 *Book back-gluing machine with drying conveyor.*

Book back-gluing machines accept books from an automatic hopper, feed, jog, glue and eject into the drying system where the books will be subjected to alternative draughts of warm and cold air. On one machine the books are glued whilst clamped but allowed to dry without any pressure (fig. 12.5). A more recent machine clamps the books between pressure bars for both gluing and drying.

Trimming

Small quantities of publishers' work may be trimmed on a single-knife guillotine, a special clamp dressing being constructed to accommodate the swell when the heads and tails are being trimmed. However, most work is passed through one or other of the types of three-knife trimmers to clean the edges and reduce the books to finished size. It is the practice in most houses to trim books to the previously prepared dummy case and it is a matter of nice judgement on the part of the operator to ensure the correct size of squares when the book is subsequently rounded and backed, lined and cased-in.

Two main types of machine are currently available, the 'striker' and the 'in-line'. The striker trimmer is front loaded by the operator, placing the work directly beneath the knives and bringing the clamp into contact by the foot-operated treadle. A hand-operated starting handle or a push-button mechanism then brings down the guard and commences the cutting cycle. When the cycle is complete the guard returns to neutral and the work off-loaded at front or rear of the machine, depending upon type. The chief advantage of this machine is the good clamping control obtained before the cutting cycle is started, but of course it is much slower than the in-line type.

The fore-edge knife of the in-line machine is positioned away from the operator and the books are delivered to the cutting position by an automatic or semi-automatic in-feed clamp (see Chapter 11). The knives operate in a totally enclosed cabinet and no moving guards are needed. The arrival of the pile of books in the trim position triggers off the clamp and trim cycle, the knives then coming to a halt. The cycle recommences with the next pile being fed in, the first pile being ejected by separate mechanical means or being thrust forward by the incoming pile. Because the operative is relieved of all tasks except placing the work in the in-feed clamp, running and production speeds are much higher and this type may be considered the standard.

The dressing of the clamp to accommodate the swell in the spine of the book contributes to the degree of accuracy obtained. Edges must be clean and mark free, particularly for the production of coloured and gilt edges. Work that has come through a calico-lining machine will almost certainly have a small overhang of material at both head and tail. The head projection may foul the head gauge on its way into the cutting position and for this work a machine equipped with a moving head gauge will be needed.

For various reasons it is seldom convenient to produce case-work two up, although this does occur in diary production. The trimming of two-up work can be effected on both types of trimmer mentioned by fitting the pile-dividing device. This allows the upper half to be trimmed and the separated bottom half to be returned to the operator for repositioning in the machine. If the work so treated has been previously glued up or calico lined the action of the tail knife will be to distort the bottom half books and even to tear the calico or other linings used. This may be acceptable on work up to 5mm thickness but the problem gets progressively worse as the books get thicker.

Edge decoration

The methods described in an earlier chapter are followed for the decoration of case-book edges. When considerable quantities have to be processed better handling methods may be adopted to speed the flow of books through the equipment. For colour spraying this may involve passing the books, in small containers, through the spray cabinet on a roller or belt conveyor.

The problem of gilding large quantities of heavy volumes is also one of handling and this is solved by utilising suitable handling equipment. The gilding press is flanked on both sides by units in which the books may be assembled, sanded and washed over with the gilding adhesive. A small powered travelling gantry above is used to lift the clamp into the gilding

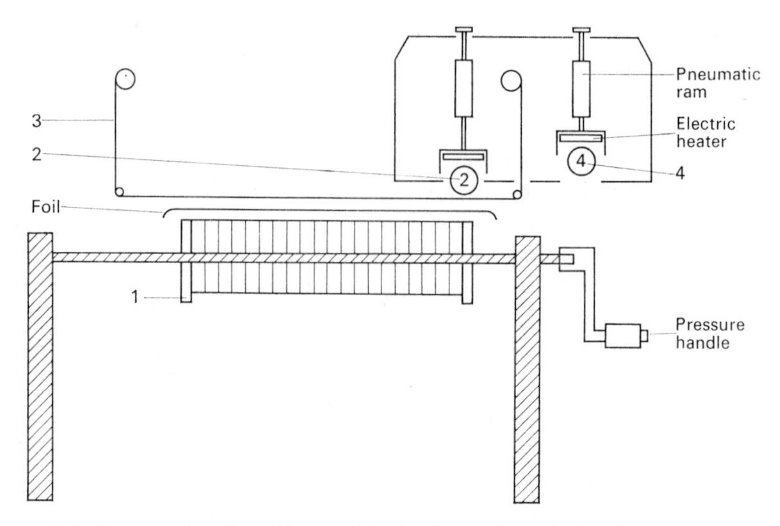

12.6 *Side elevation of a roller-gliding machine. Books are clamped* (1) *and sanded* (*not shown*). *After preparation with a thermoplastic fluid and allowed to dry, the ribbon gold or foil is laid on. The head is then moved to its operating position and roller* (2) *works through a rubber blanket* (3) *at about 290°C and preheats the surface. Roller* (4) *follows, working at about 320°C, reactivates the adhesive and ensures the transference of the coating from the foil to the bookedge.*

machine for actually applying the foil by heat transference. The assembly may be lifted back for opening and inspection. In this way the press is used to fuller capacity (fig. 12.6).

Some attempt has been made to automate the gilding process but this has not been taken up by the industry. A machine put on exhibition some time ago had a continuous belt clamp into which the books were fed and transported over various stations to sand, prepare, dry and gild using a continuous roll of foil.

The success of the gilding operation relies very much on the building up of an experience log in which the variable factors are recorded. Suitable heading for such a log are title of book; nature of paper, if possible with sample; adhesive preparation used (an alternative to shellac is a watered-down solution of polyvinyl acetate adhesive); pneumatic pressure reading; temperature for both the direct and indirect rollers.

Rounding and backing

Machines for rounding and backing may be classified into semi-automatic and automatic versions and both have a place in the bindery. The older semi-automatic types, although relatively slow in production, have a very wide size range and this makes them almost indispensable in the production of both very large and small books. The glued-up book is laid on to a table and the spine held against two concave gauges whilst the rounding rollers grip the book. From here onward the machine function is automatic, the rollers applying pressure and drawing the book into the required shape. Control of the rotational action of the rollers gives more or less round as required. Rigid jaws now grip the book, holding it in shape as the rollers relinquish their hold. These carry the book in an arc to the backing position, where the selected wiper or backing plate shapes the spine by two movements across the spine in both directions. A tension bar operates during the second cycle of the machine to ensure that a really sharp shoulder is produced. The jaws now return the book to the feed position where it may be hand removed or transferred to an automatic delivery device.

Limitations on this type of machine derive from the variations that occur due to hand feeding and the design, which allows only one book to be processed at a time.

Automatic machines are constructed on the flow-line principle, several books being in the machine simultaneously. Books are placed in the feed trough, fore-edge down, and are then automatically transferred to the succeeding stations at which rounding, backing and delivery occur. A hopper feed may be fitted and delivery can be directed to a transfer conveyor passing the books to the next process as part of a linked system. Running speeds of around thirty books per minute are usual although this may be reduced to manage very thick or difficult work (fig. 12.7).

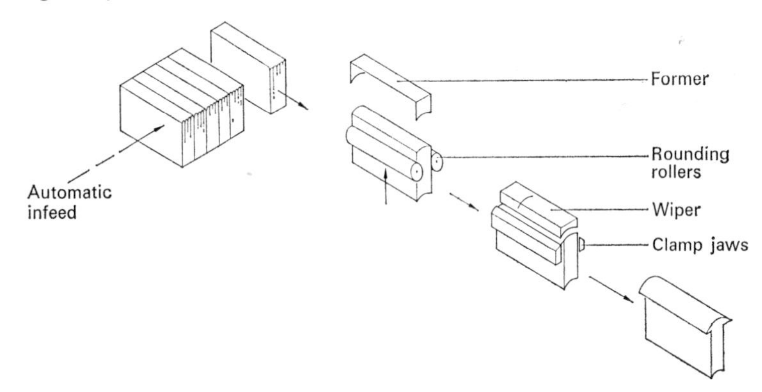

12.7 *In-line rounding and backing process.*

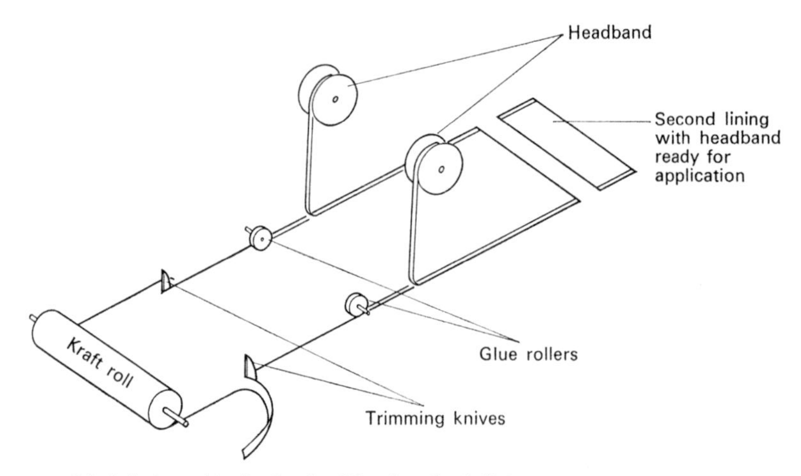

12.8 *Method of attaching head and tail bands to kraft lining.*

Another type of rounding and backing machine that has appeared in recent years completes both the rounding and backing action by the use of hydraulic pressure. Although not strictly a high-speed machine, the ability to switch from one thickness to another without altering the machine controls makes it a very useful addition to the bindery handling short runs and sizes outside the usual range. Operating speeds of around ten books per minute are claimed.

For all rounding and backing operations it is important to see that the books have not been spine-glued too far in advance; a period of two hours may be considered as average. When books are left overlong the resistance of hard glue may preclude good shaping or accurate shoulders. Operators frequently wipe over the spines with a grease dolly to ease the movement of the backing plate over the back and to prevent any sticking of the book in the jaws.

Lining

Conventional sewn case-books usually have a first lining of mull and a second lining of crêped kraft paper. Some machines incorporate the facility of applying two mull linings and also head and tail bands. A double mull is useful when large heavy books have to be reinforced for hard wear but extra care is necessary to ensure adhesion of the mull on endpapers at the casing-in stage; to reinforce spine only the first layer of mull may be cut to spine width. Decorative head and tail bands may be of silk, cotton or twisted rayon and these are attached simultaneously with the second lining (fig. 12.8).

Probably the most widely used machine has a continuous chain of spring-loaded clamps (fig. 12.9). On its lower level a clamp opens and a book is placed in position; the clamp then grips the book and the chain moves along one station. In this manner the book traverses the machine, the necessary glue and linings being applied at intervals along the machine. Positioning the

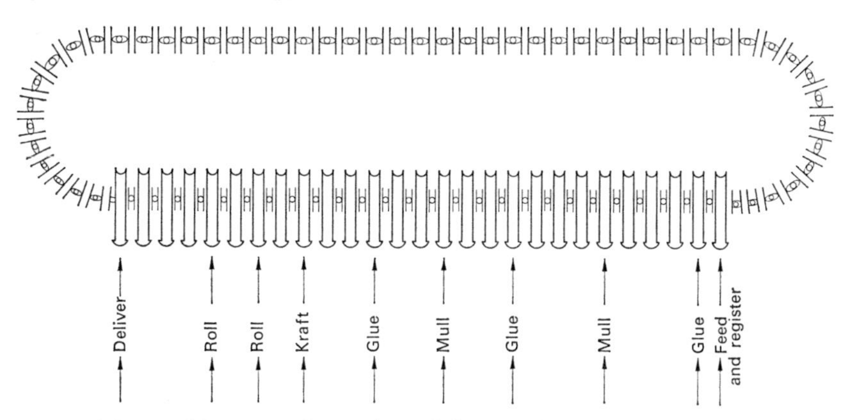

12.9 *Side view of the conveyor clamps of a triple-lining machine.*

book correctly in the feed clamp is important if good gluing is to be achieved at later stations and a former is used for this. Glue is applied lengthwise from a shaped concave rubber roller and this is flexible to accommodate to any slight variations in the book shape. Strips of both mull and kraft are cut from a roll that has its width the approximate book length. This means that in this instance the machine direction of the kraft and the warp of the mull lining are both across the book spine which is contrary to the principle set out in part one of this book.

There are blank stations between various components, *eg* between gluing and mull attachment, and this allows the adhesive to become tacky and ensures instant adhesion when the lining is dabbed on. The headbanding attachment glues the fabric headband material to the edge of the kraft web as it is fed to the machine and when the appropriate strip is cut off and attached the whole assembly is dabbed into position. It is important that the width is correctly gauged if the fabric headband is not to crumple when the case is drawn around the book.

The final action on the book before delivery is a roll over the spine with wetted pneumatic rubber rollers. The soft rollers take up the convex shape of the spine and thus ensures reaching every part of the lining; the moisture applied softens the kraft and allows it to shape readily to the book. Later this lining will shrink on the back to give a snug tight fitting.

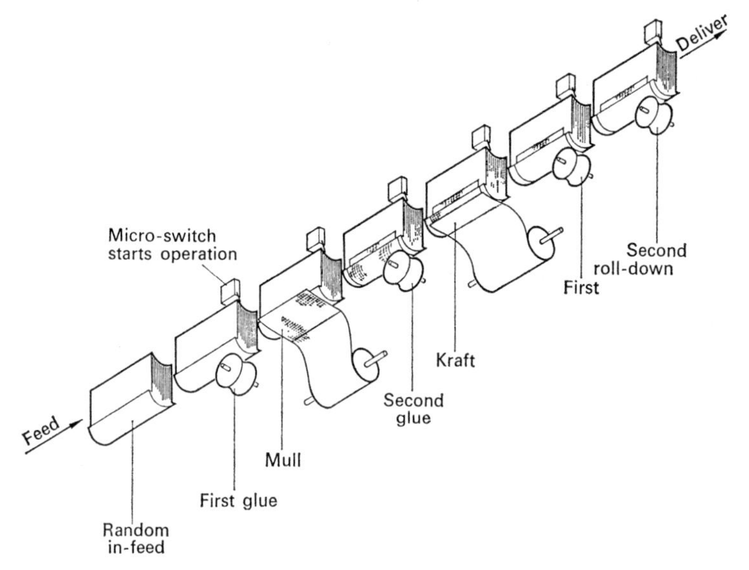

12.10 *New style lining machine.*

The delivery on the machine is designed to allow the book to be removed with a minimum of disturbance of the glued linings and book shape; transference to the next process may be by hand or direct by specialist conveyor.

A group of lining machines recently offered process the books spine downwards in a continuous stream, head to tail, and gripped by canvas or chain belts. Located beneath the chain are gluing and lining heads in various combinations. These are actuated by solenoid switches which are depressed by the approaching book (fig. 12.10). No particular time cycle is observed in the feeding of this type of machine and running speeds are relative to the length of the book and the speed of the clamp belt. A feature of these machines is the ability to process long, thick and landscape work. For quantity production the machine may be linked with rounding and backing and the casing-in machine by specialist conveyor.

Alternative forwarding techniques

There are two regularly used alternatives to the previous binding sequence, one based on a sewn book and the other on adhesive binding. The same type of machine may be used for both techniques and these may be described as calico lining machines (fig. 4.24.1). A sewn book fed into the canvas belts is clamped and glued. A reel of lining material is also glued and the book attached to it in a continuous stream, the extra width of the lining being turned over on to the endpapers and rubbed down. Just before delivery a sharp knife enters the 20mm gap between books and separates each book in turn and these are then stacked for drying.

For the production of adhesive bound and lined books, gathered and unsewn sections are fed, the spines milled, glued and lined as before. Many of the comments made about adhesive binding in an earlier chapter apply here too.

These styles are very suitable for square backed work of the cheaper variety but when applied to rounded and backed work on heavy stock little support is given to the sections. For this reason both sewn and adhesive bound versions tend rapidly to lose any shape imparted by the rounding and backing process. Another hazard here is the risk of the lining material fracturing during the backing operation leaving the spine weakened along a line head to tail of the book. Because of the ability of these machines to process non-standard work they are often used for large work that is then second lined with kraft paper as a hand operation.

Any suitable lining material may be used for the spines of square backed work, thin dressed fabrics and long fibred papers being best. Subsequent rounding and backing requires that the material should expand with the book during this process; stretch cloth or expanding calico, latex and plastic impregnated papers are suitable. When calico lining is used a change in the production sequence is necessary, *eg*

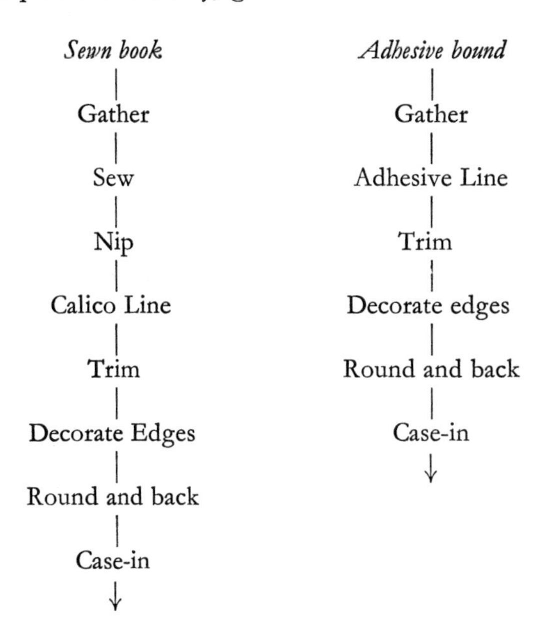

It should be noted that this type of lining machine is very suitable for the lining of other classes of work when large quantities are being handled. Such work includes quarter bound cut flush memo books, reporters' notebooks etc; duplicate and triplicate order and receipt books and computer print-outs.

Some of the larger multi-clamp adhesive-binding machines that normally produce wrappered work may be equipped to fulfil the dual role of lining for adhesive-bound hard-back books. The usual method here is to use a layer of mull under a layer of crêped kraft paper, a side gluing attachment ensures that these materials adhere to the endpaper.

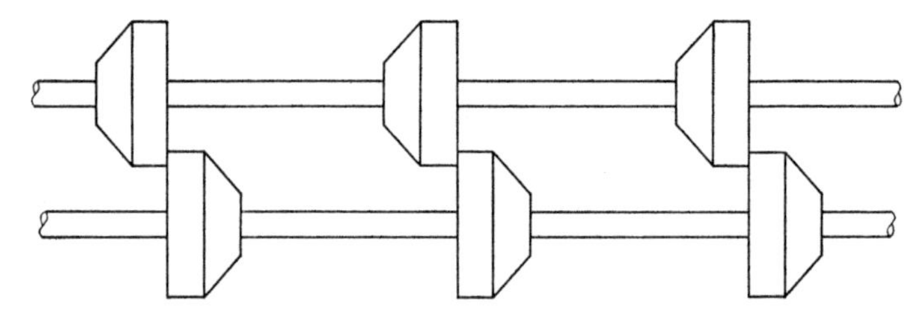

12.11 *Rotary sheer cutters.*

Board cutting

Book boards are reduced to size from large sheets by a rotary shear action machine (fig. 12.11). Two runs through a standard machine are needed to reduce the board to the required size, one to produce strips from the large sheet and the second to complete the cross cut rectangles. Accuracy of cut is obtained by ensuring even pressure on both the feed and delivery pressure rollers. Most machines have the facility of automatic hopper feeding of the strips. The push feed setting allows faster or slower movement according to the length of the cut in relation to the circumference of the cutters, *eg*

3 feed movements to 1 rotation of the cutters
2 feed movements to 1 rotation of the cutters
1 feed movement to 1 rotation of the cutters
1 feed movement to 2 rotation of the cutters

Thus a strip having a small dimension of 250mm may be cut on the second setting if the circumference of the cutter is 800mm.

Cutting boards on the rotary shears gives a relatively 'soft' edge to the board in comparison to that produced by a guillotine cut. This tends to be advantageous as sharp corners may wear through some bookcover materials in use. Producing book boards on the guillotine has other disadvantages too;

the guillotine is a more expensive piece of machinery; the knife rapidly blunts and slight variations in size between top and bottom boards results in minor variations in case size; and double dressing of the offcut is necessary to remove the bevel edge.

Modern rotary board shears have powerful suction feed units that can withdraw a large full-size board from the bottom of a 700 kilo stack and feed it into the cutters. Delivery is on to a travelling belt with automatic stacking at the end. One of these machines may produce 100 000 pairs of boards per week.

L-shaped machines have appeared in the industry from time to time but have seldom found favour. These have two cutter units arranged at right angles with a redirectional device between.

As a general rule machine direction of board should run parallel to the spine but for economy reasons this is often sacrificed to get an extra board or two from the sheet. Occasionally an extra board may be produced by 'stagger' cutting but as this requires an extra setting of the machine it is important that the saving in material is not offset by the extra labour cost. Usually it is undesirable to attempt stagger cutting unless a long run is involved.

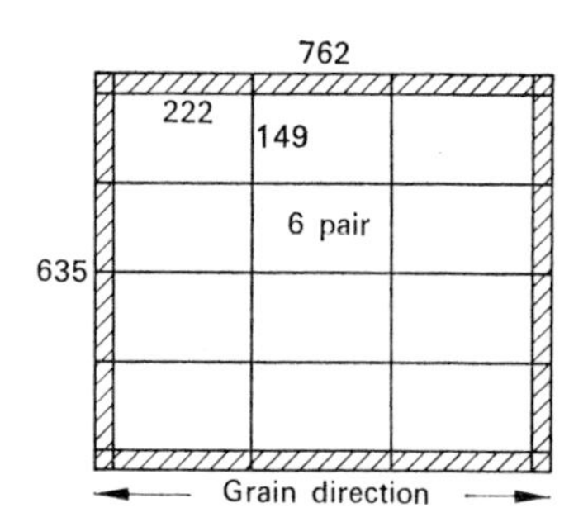

12.12.1 *Cutting boards—grain correct.*

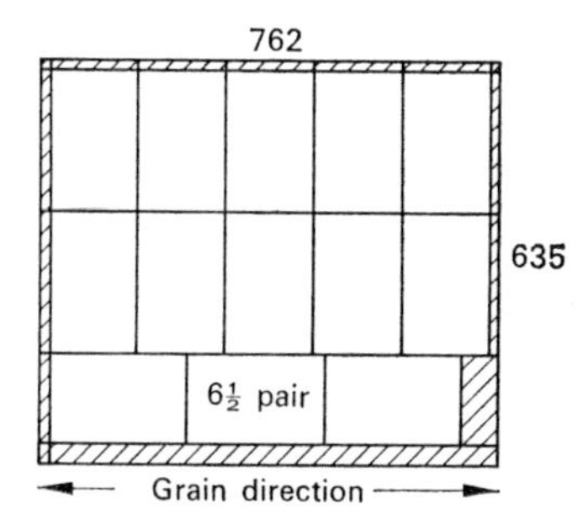

12.12.2 *Cutting boards—economical cutting.*

Quality of the finished boards is checked on delivery by knocking up and rejecting those that are out of square. A simple method of checking squareness is to reverse two boards; any error will be doubled and immediately apparent. Boards that are to become part of a round cornered case may now be shaped in a hand or pneumatically-operated round-cornering machine.

Two main areas of difficulty that occur in the use of chip and strawboard are concerned with lamination and maturing. Failure of the bond between the substrates is clearly a manufacturer's problem and little can be done except return the batch to its source. Immature boards, sometimes called 'green', show various forms of distortion when shipped directly from the mill in, perhaps, wet and cold conditions, and are put immediately into the production line in a warm and dry workshop. The simplest method of curing this fault, and one that is adopted by many book manufacturers, is to carry sufficient stock of boards to allow a few months for the board to stabilise in reasonable conditions before they are used. If this is done the strappings or tyings should be released and the pile restacked occasionally to allow the stabilisation to occur evenly throughout the stack.

Cloth cutting

The preparation of the covering material for the casemaking machine may mean slitting and cutting into rectangles of the required size or slitting and rewinding the reel for a reel-fed casemaker.

Cutting off rectangles by hand methods and then trimming round on the guillotine is wasteful and costly; a saving of up to 15 per cent may be achieved by using a machine. When mounted on the machine the web of cloth is passed through slitting knives that divide the reel along its length. A friction-driven roller is set to eject a given length of material before severing it from the reel by a dip-shear knife (fig. 12.13).

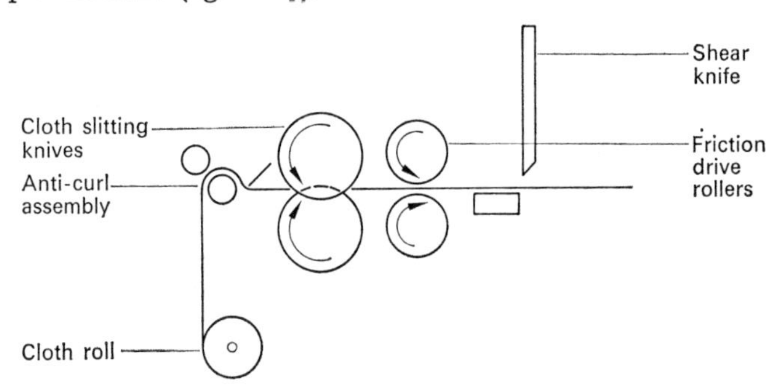

12.13 *Principle of the cloth cutting machine.*

Although it is ideal to cut woven fabrics with the warp and the grain of paper materials parallel with the spine, it is often necessary to disregard this for reasons of economy. Occasionally extra saving is made by stagger cutting and this is possible on the cloth cutter by withdrawing and re-reeling the odd dimension. This is later mounted in the feed position, the 'throw' reset and suitable lengths chopped off without using the slitters. Cloth with warp the short way of the sheet will react quite differently on the casemaking machine to that with long warp and should be processed separately.

Woven fabrics all have a selvedge which is much thicker than the rest of the cloth and this prevents the reel from having nicely aligned edges. During the cutting of a size that comes within 10mm of the selvedge it may be necessary to constantly realign the reel to prevent the cloth from running out of the end slitters. Difficulty may be found in cutting very soft unstiffened materials such as soft canvas and unsupported polyvinyl chloride, very thick substances *eg* thicker leathercloth and those materials that have remained on the reel for a considerable time; the latter often shows as a variation in the length of the throw. As the material should be as flat as possible when used on the casemaker an anti-curl device is fitted to the cloth cutter and this is more or less successful according to the type of material being cut.

The machine used to produce reels for the reel-fed casemaker is a slitter-rewinder. The rewind mechanism has two shafts so that adjacent webs are reeled on to alternate shafts and these short reels are later joined into composite reels of about 500mm diameter.

Casemaking

The manufacture of small quantities of publishers' cases is completed by hand using bench methods or assembly jigs. A typical line of small machines used in the latter instance includes sheet gluing machine, case assembly jig, turn-in machine and rubber roller press.

A sheet-fed casemaking machine may be hand or automatically fed, cloth feed and gluing at one end and board and hollow feed at the other end of the machine. A sheet of cloth is taken from the top of the pile by suckers, placed on the gluing cylinder, gripped, glued and delivered to the assembly area in the centre. Boards are pushed from the hopper into slides and a piece of hollow paper fed from a reel and cut to size. A board transporter lifts the boards and hollow in correct position and places them centrally on to the glued cloth. With the transporter maintaining its position, folding bars turn the head and tail turn-in over so that it adheres to the inside board; simultaneously corner 'tucking' devices correctly tuck in the corners to obtain a neat effect. Still under pressure from the transporter the platform and the case is lowered to another level at which the fore-edge turn-in bars operate thus completing the case. The transporter moves away for the next cycle and the case is delivered to the press where it remains for one cycle of the machine before ejection into the delivery rollers. This standard type of sheet-fed casemaker has control switches that will stop the machine in the event of malfunctioning and will produce between 5 000 and 7000 cases in an eight hour period (fig. 12.14).

A recent advance in the design of these machines has produced the so called 'double head' version. This replaces the reciprocal motion of the single trans-

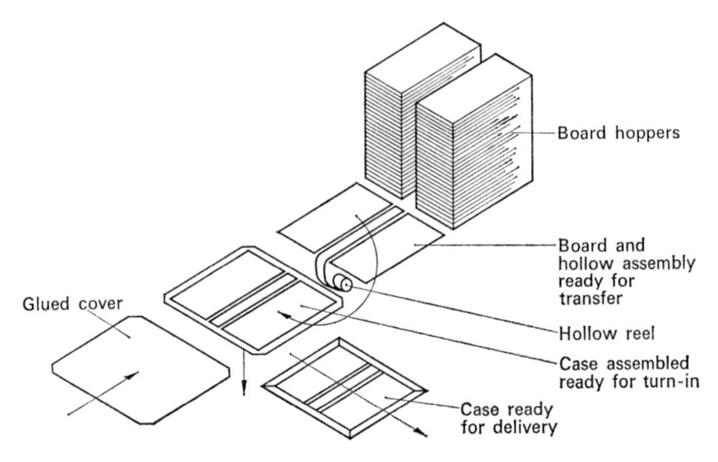

12.14 *Stages in sheet-fed casemaking.*

porter with a twin head having a rotary motion. Other design improvements, including the gluing mechanism, allow for running speed of 30 cases per minute and a net production of between 9000 and 10000 cases per eight hour day.

The product of sheet-fed machines includes quarter bound work but this requires a double run; the spine cover being attached on the first run and sides on the second time through. Round corners are handled by the machine but most manufacturers subject each corner to a light individual pressing after it leaves the machine. This is a treadle or pneumatically operated device having an exchangeable bed surface, for grained work, and a plunger head that sharply taps home the surplus cover material, shaping up the corners and ensuring adhesion. When pre-printed covers are being made it is essential to obtain good register of the printing in relation to the boards and hollow; various devices on the machine enable this to be done. Speed of production is to some extent governed by the initial tack of the adhesive used and for this and economy reasons animal adhesives are used exclusively.

Reel-fed machines are less popular than the sheet-fed varieties probably because of the higher capital cost and the longer set-up time. The web of cover material is fed off the reel, over a gluing mechanism and on to the assembly table. As the cloth moves along station by station, boards and hollow are added, V cuts made for corners and the fore-edges turned over. A case is then detached from the web, the remaining two edges turned in and the case delivered through rubbing down rollers. Quarter-bound cases are made by feeding three separate reels of material often with the spine material overlapping the sidings; this is a useful recognition factor because it is contrary to the sheet fed practice (fig. 12.15).

Special problems arise during the manufacture of diary and other loose sided covers. This type of case requires that only the turn-in of the cover is adhering to the thin caliper cover board. This imparts a very flexible feel to the cover and in pocket books allows them to mould to the shape of the

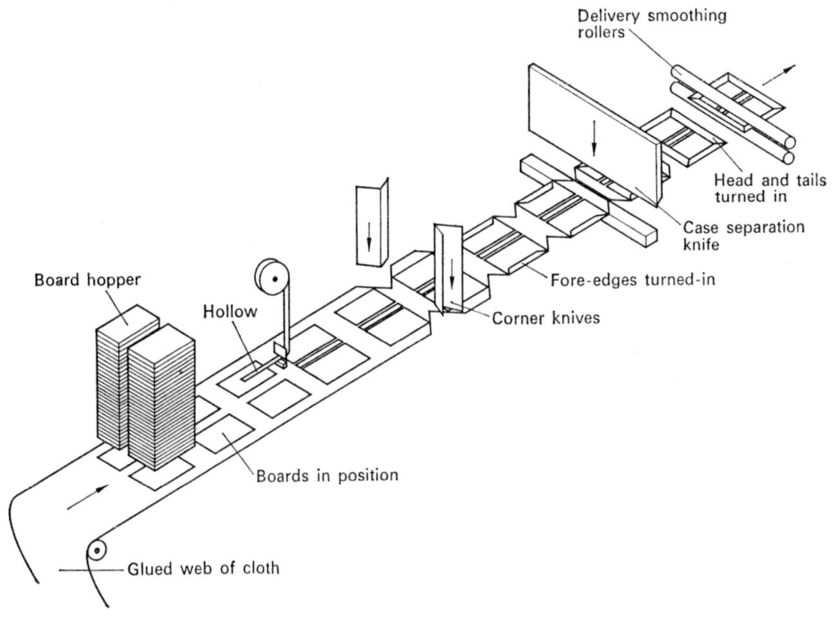

12.15 *Stages in reel-fed casemaking.*

body, when in use. One solution to this problem is to be found on some sheet-fed casemakers which have the normal glue applicator assembly replaced with the stencil-gluing attachment. Although the set-up time is longer than usual this principle can be used successfully on a variety of covering jobs other than cases and these include silk covered pads for greetings cards with or without foam filling and display material. To feed very thin boards on a sheet-fed casemaker it is necessary to replace the traditional push feed mechanism with a device that closely resembles a paper feeder found on some printing machines. This is wheeled up to the machine when required and operates by placing the pair of boards directly in position on the board in-feed rails. In many instances boards of diaries are stamped out using a cutting and creasing forme, the two boards remaining joined by a perforated spacer until after the case is made, when it is torn out.

Cases made with very thin boards and having rounded corners cannot be made on the traditional casemaking machine. A method of producing this class of work requires that the covers are coated with thermoplastic adhesive and dried. The centre point of the casemaking machine used is a heated former, the aperture of which is exactly the shape of the case to be made. The case materials are loaded into the various holders and the head closed so that the male unit engages into the female former. The cover is turned-in, the corners pleated and sealed down by heat reactivation in a few seconds. The assembly is then removed and allowed to cool under pressure.

A popular innovation in children's toy books and other square-back case-books require that the cover has a stiff board hollow. Both reel fed and sheet fed machines can be adjusted to feed the board hollow which is housed in a specially made hopper and positioned between the normal board hoppers.

Cover decoration

The method chosen to title and decorate the cases of publishers' work depend to a large extent upon the use and cost of the volume and may include blocking in ink and foil, printing by offset, letterpress or screen process before or after casemaking and labelling. The use of offset pre-printed covers on a paperfelt stock has made great strides in recent years and a high percentage of educational textbooks and primers are now produced this way. Printing is a very economic method of production but it must be recognised that the surface printing will have a limited life, the ink being easily scuffed by everyday handling. Combinations of pre-printed materials and foil blocking are often used in the production of prestige and presentation books of good quality. The use of other methods of inked cases has declined and this applies equally to letterpress and to blocked work. Blocking is an attractive and permanent method of marking cases utilising both metallic or pigmented foils.

Although most wholesale bookbinders use a hand blocking machine for very short runs and dummy copies, most of the productive work is completed on one or other of the power-operated machines. These may be divided into three main groups and described as having vertical acting, platen and rotary impression principles. The first of these is the traditional type and is sometimes called a 'pillar press' because of the two or four round steel pillars that link the head to the base to give rigidity; another version of the same type has a wide inverted 'U' brace over the top of the machine and these are called 'horse-shoe' presses. The four-pillar press has great rigidity in its action and tremendous pressures may be exerted. This makes the machine very useful for the heavy impressions sometimes needed for 'blind' decorative work and for heavy blocking on deeply grained surfaces, *eg* leathercloth and art canvas.

Most of these machines are hand-fed although automatics are available feeding from off the bottom of a case hopper at speeds around 40 cases per minute. Another modern version of this type has pressure control, variable timed dwell of up to six seconds, variable speed, magnetic clutch-brake mechanism and a control system to give a variety of operational possibilities on the platen movement, *eg* 1, single operation and stop; 2, continuous run; 3, timed dwell on impression with continuous feed; 4, continuous run on impression with timed feed; and 5, timed feed and timed dwell. With this control over the functions of the machine a wide range of difficult work can be satisfactorily handled, *eg* soft leather covers for graining or preblocking, silks, etc, using controls 4 or 5; very heavy impressions on work easy to feed using control 3, etc.

Although ink blocking has gone out of favour some heavy platens are still used for foil blocking and one version is particularly useful for casework. This utilises a feeding device into which the cases are loaded standing on edge. While feeding is going on from the front of the feedboard, loading may be continued from the rear and in this way long continuous runs are possible (fig. 12.16). Other heavy platens with conventional feeders are less successful

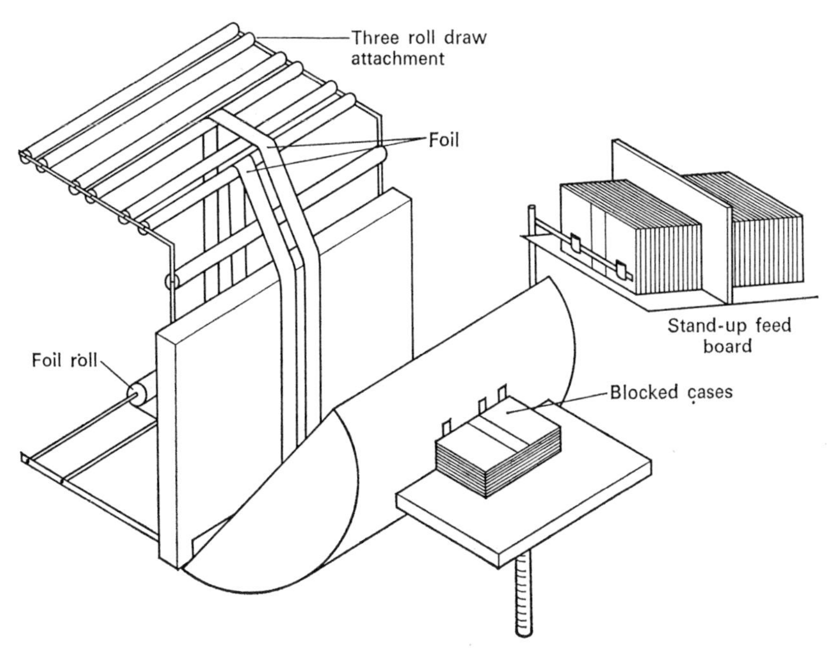

12.16 *Heavy platen equipped for foil blocking with heavy duty feeder.*

on caseblocking but can achieve very high outputs on thinner materials such as magazine covers, carton blanks, greetings cards, etc.

A third type of machine used for very long runs utilises the rotary principle (fig. 12.17). A curved engraved relief block is fixed to the plate cylinder and cases are fed from the bottom of a hopper into the nip between plate and impression cylinders. Machines are equipped for both ink and foil work and are particularly useful for blanking out the blocking area with a crushing brass for subsequent foil titling. Production speeds on foil work are considerably lower than those reached on ink blocking but are still extremely high in relation to conventional machines.

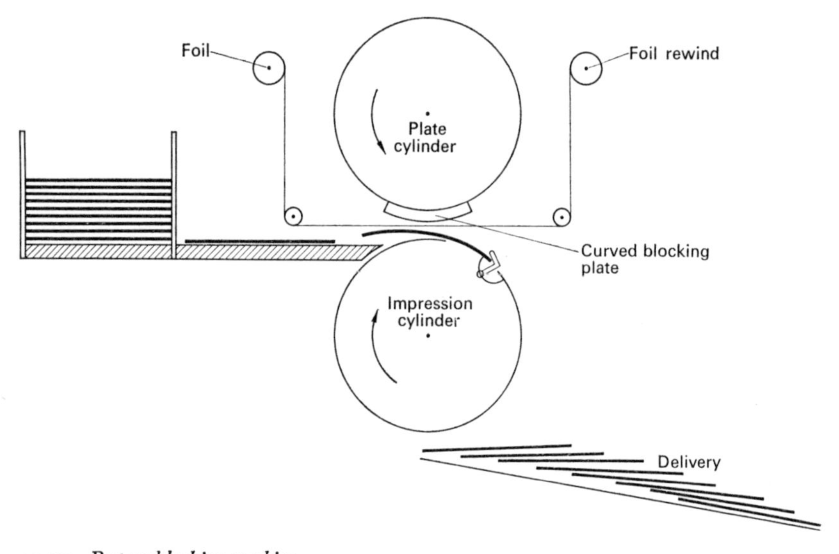

12.17 *Rotary blocking machine.*

Foil-feeding attachments are banks of rollers in pairs, each pair operating separately and being set to pull through different lengths of foil (fig. 12.18). These are usually positioned at the rear of the blocking machine so that foil must run from front to back. There are some advantages to be gained if the foil can be fed across the width of the machine and some four pillar presses have this facility. In some instances rewind attachments are also fitted and these can be of great assistance in the economical use of foils. Most blocking problems are related to the nature of the surface receiving the impression, the qualities of the foil used and the various permutations of the three principles of blocking, *ie* heat, pressure and dwell. Many binding cover materials are now surfaced with polyvinyl chloride, cellulose and other spray finishes and

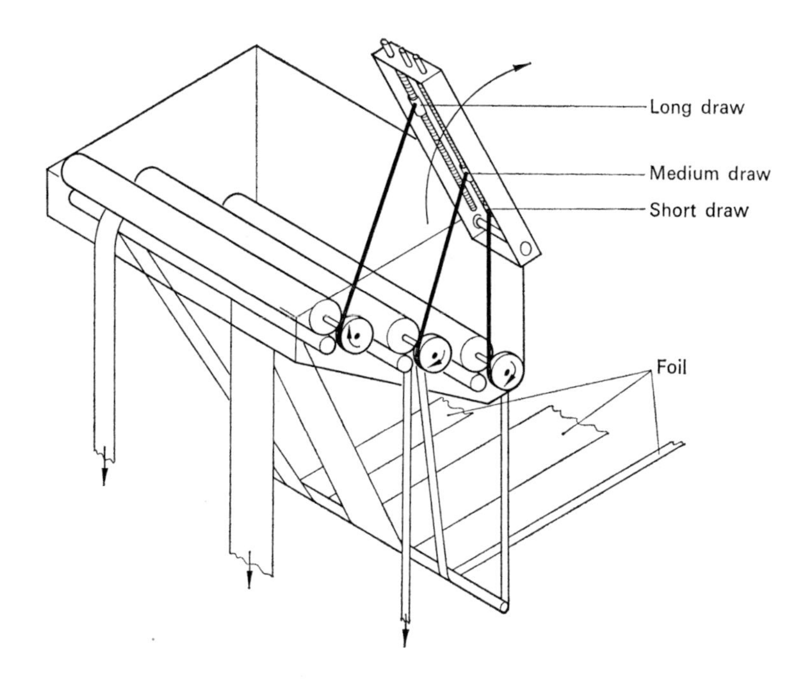

12.18 *Three-roll foil feeder showing each roll set to draw different length.*

the size used on foil may be incompatible with this surface. Manufacturers recommend that a trial run be blocked off every time a new material or foil is introduced into production and some offer two quite distinct qualities; one suitable for fibrous materials such as paper, board and cloth and another for plastic surfaces.

The blocking of delicate plastic surfaces will require great care in machine setting. As the surface is thermoplastic a few degrees above that necessary for reactivating the size may melt or blister the PVC. Depth of impression is also critical; because of its elastic nature direct heat blocking will not bottom the grain and the surface will return to its original shape when impression comes off. In certain circumstances grains in plastic materials can be flattened by the use of a short high frequency charge working through a shaped brass block; this blind impression can then be foil blocked with colour or metal. Little trouble will normally be found in bottoming the surface grains on most binding materials although the very heavy grains of some leathercloth, leathers and canvas materials may need considerable dwell to achieve the right result.

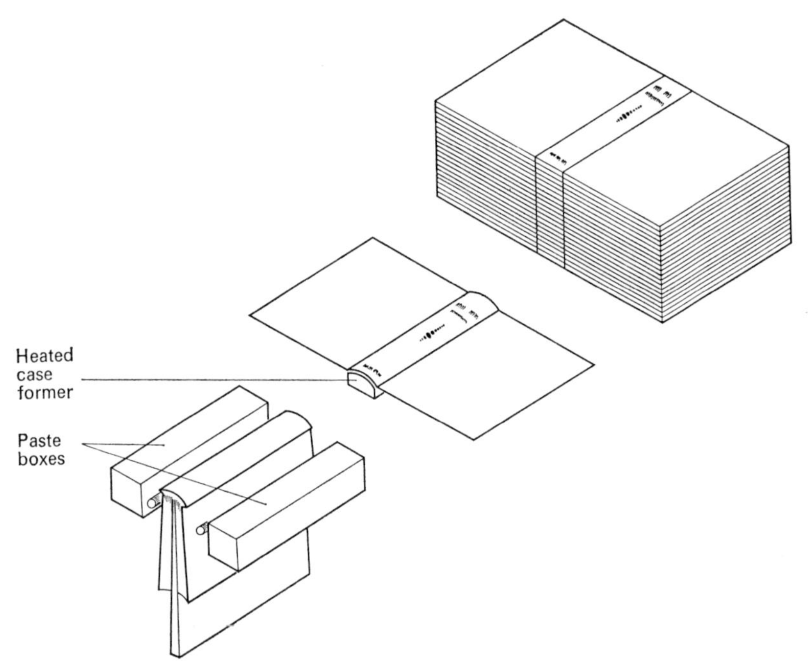

12.19 *Stages in casing-in.*

Blocking foils appear to have a limited shelf life and if foils fail to adhere, deterioration of the size may be a contributory factor.

Casing-in

Joining the book to its cover is an operation that must be completed successfully if the book is to have a reasonable life expectancy. Cases are loaded on to a modern machine and ejected from the hopper into a case rounding device. This consists of a heated semi-elliptical bar and forming arms to keep the case in firm contact with the heated surface. By careful selection of the former bar and the setting of the arms the spine of the case is correctly shaped to fit the book and it is then passed to assembly area of the machine. Here the book is centred on an arm and lowered between two pasting boxes. The spring-loaded boxes move into contact with the endpapers and as the book rises on the arm a continuous application of paste adhesive is rolled on to the surface of the endpaper. Most paste boxes have some mechanism to ensure extra adhesive on the mull lining and into the shoulder. As the case is positioned above the arm it is picked up and correctly positioned on the rising book. A pressing clamp lightly sets the case into position before delivery.

Machines having one, three and six arms are available, their application depending upon the type of work to be completed. The single arm type is largely used for short runs of square back books or those that have been rounded only, both books and cases being hand loaded and off-loaded (fig. 12.19). Machines with three arms working in a horizontal manner are hand fed at the first position, pasted at the second position and off-loaded at the third. Because of the so-called 'half turn' device incorporated into the pasting mechanism this machine will successfully apply adhesive to the endpapers of rounded and backed books. The multi-arm machine has a continuous vertical motion, the book being picked up as the arm rises and being carried through pasting, case attachment and setting before the book is removed from the arm so that it may return to the bottom of the machine to repeat the cycle (fig. 12.20).

Successful casing-in depends largely upon the quality of the operations which have gone before, so that book and case marry up without the necessity

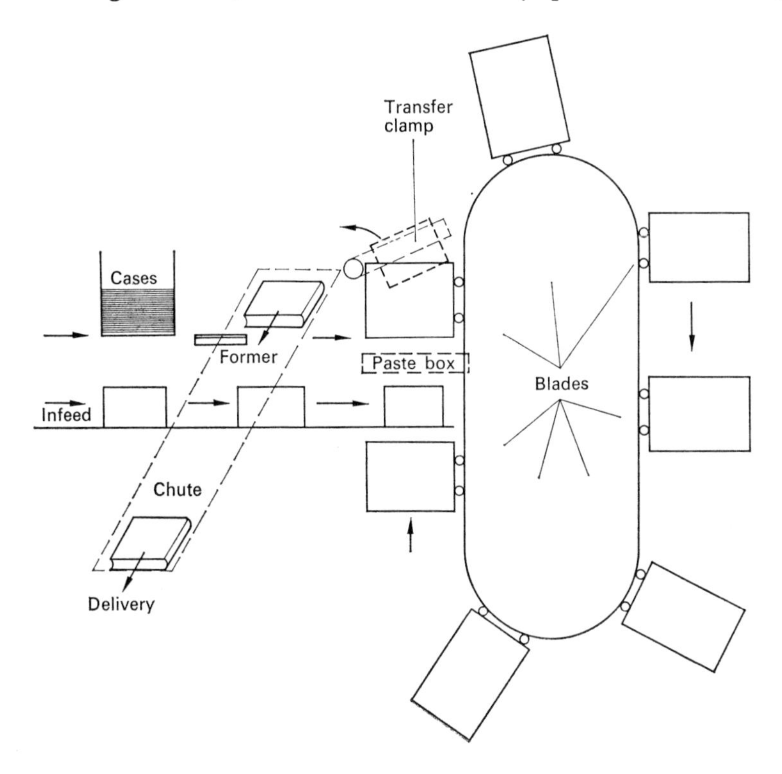

12.20 *Side elevation of a continuous casing-in machine.*

of hand adjustment of the squares. The high speed continuous machines often pass the books directly to the next process and the opportunity to adjust the fit of book and case does not exist. Work that is thick, has a large area or for some other reason cannot be processed on the continuous machine will be cased-in on the slower three-arm machine. The quantity of adhesive in the shoulder governs the success of the join but excessive adhesive on the remaining area of the endpaper may lead to it oozing or being squeezed on to the book edge and sealing off a number of leaves as it dries.

Thinned starch paste is the adhesive favoured for casing-in as it penetrates deeply into the fibres and gives a good bond between endpaper and boards. It also has a long open time which gives a good margin of time for the book to be assembled in a standing press. Where the book goes directly to a book forming and pressing machine then a paste modified with PVA is normally used or a joint gluing attachment may apply a narrow strip of PVA in the shoulder before it enters the casing-in machine. PVA is used because the heaters on the book forming and pressing machine can quickly dry off the adhesive during its short passage through the machine. This technique is particularly valuable in preventing joint failure on books that are lined with calico or similar materials.

Another problem that occurs during casing-in is that of adhesion between the endpaper and the turn-in of the case when the cover material is a nitrocellulose or PVC impregnated surface. Being impervious these materials will prevent mechanical adhesion and it is necessary for the starch paste to be modified with a substance that is sympathetic to these surfaces. In practice this again means the addition of certain formulations of PVA to the starch and some adhesive manufacturers offer, as standard, a casing-in adhesive which will satisfy all the conditions above. However, the cost often precludes its use as standard and it is only used when needed.

Pressing

The traditional method of pressing books after casing-in is to build them up in a pneumatically-operated standing press. This has about 1·5m of vertical pressing space with the base set flush into the workshop floor. This allows the books to be built up on metal skids and when a container is complete it is wheeled out for the extended drying period necessary (fig. 12.21). Locking the top plate to the skid to complete the container is effected by chains or steel bars on the short edges; these are put into position whilst the head is under pressure and when the pressure is relaxed the strain is taken up on the ties. This method of pressing necessitates a minimum drying period of about two hours whilst the adhesive sets and the excess moisture is absorbed into the

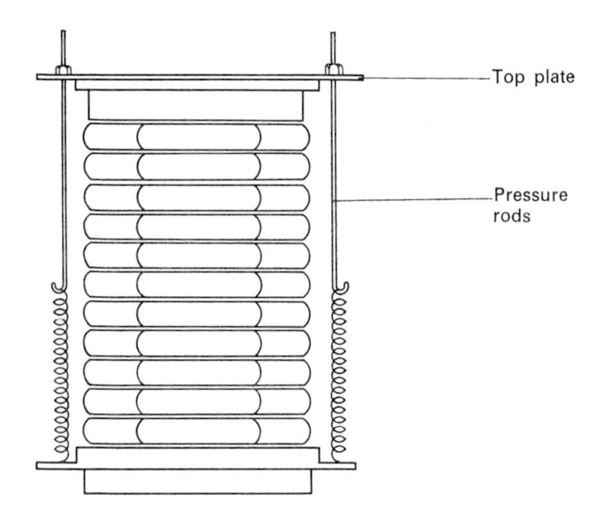

12.21 *Container of books for drying.*

fibrous materials. Books can be given longer drying time and in this way some maturing will occur, the construction being consolidated and set into shape. This is particularly important for large heavy volumes, those having heavy cover stocks and when metal-edged boards are used to press in the groove.

Assembling books into containers in this way and then dis-assembling them after the drying period is obviously a laborious and expensive task and for this reason machine manufacturers have in recent years introduced book forming and pressing machines. The intention here is to mechanise the pressing function to keep it in step with other processes. Each machine has four or more pressing stations in which the book is shaped, pressed and dried. In some versions the book is passed from station to station changing clamps on each occasion and in others the clamps move with the books. Advantages are claimed for both these methods but both types have heated formers that cure the adhesive in the joint. To enable this heat to reach the adhesive it is necessary that all books using this system have a 'semi-groove' (fig. 12.22). The original version of this machine processed the work in a straight line but probably the most popular type now made has a carousel type of action that occupies much less space.

Time spent in a machine like this is about 10–15 seconds so production is continuous with the casing-in function but no maturing has been effected. It is therefore necessary that books be given the opportunity to stabilise with atmosphere and in most binderies the books are stacked on to pallets and

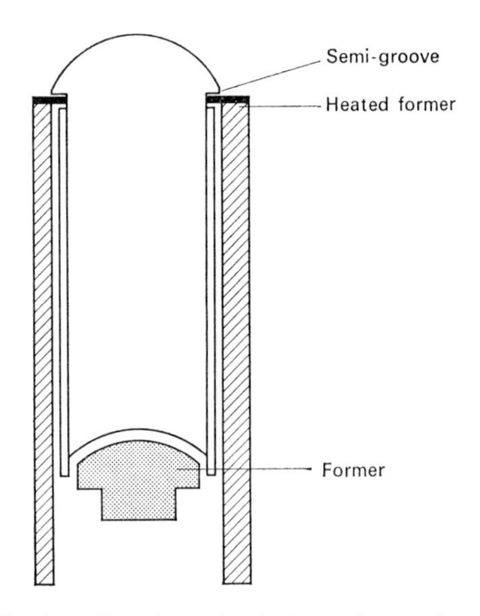

12.22 *Cross-section of book forming and pressing station showing semi-groove, heated joint former and book shape former.*

weighted to prevent any unwanted warping. The stack may then be left until required for inspection and jacketing.

Inspection

Although a certain amount of quality control is exercised during production it is usually desirable to have one last check on book quality before it receives its jacket and is despatched. The standard of this inspection may vary extensively from company to company and between various price levels but a typical standard is as follows. Run through text pages for dog ears, tears and grease marks, check end papers and title page for cleanliness and opening; inspect cover for right way up, blocking standards and cleanliness. Some companies have carried out experiments to test the need for inspection and have concluded that as the percentage returns are similar in both inspected and non-inspected output this operation can be abandoned for the cheapest work.

The visual inspection of completed work is usually linked with the fitting of the jacket by hand. This is trimmed to fit the cover accurately; oversize jackets tear easily, particularly in bookshops, while undersize jackets show an unwanted line of bookcloth. It is drawn around the book by hand, registered and the flanges tucked in.

A book-jacketing machine can do this task at quite high speed, the turn-in bars being heated to give a crisp sharp fold and good register.

Automation

Completely automatic, untouched by hand, bookbinding is a long way off although partially linked systems are available (figs. 12.23.1.2). The difficulty in engineering a system in which the book flows from stage to stage without halting derives from the nature of the units being handled at each stage, *eg* sheets at folding, sections at gathering and sewing and books thereafter. It is not really until the book reaches this latter stage that it becomes possible to consider production in unitary form. Previous to this the book may contain from three to sixty or more sections and will require this number of cycles in the processing machinery. These facts are illustrated by considering two jobs of 1 000 books; job A has a make-up of 20 × 16-pp sections job B is made up as 10 × 32-pp sections. Both these jobs may have identical times in the folding and forwarding stages but the sewing of job A will take twice as long as

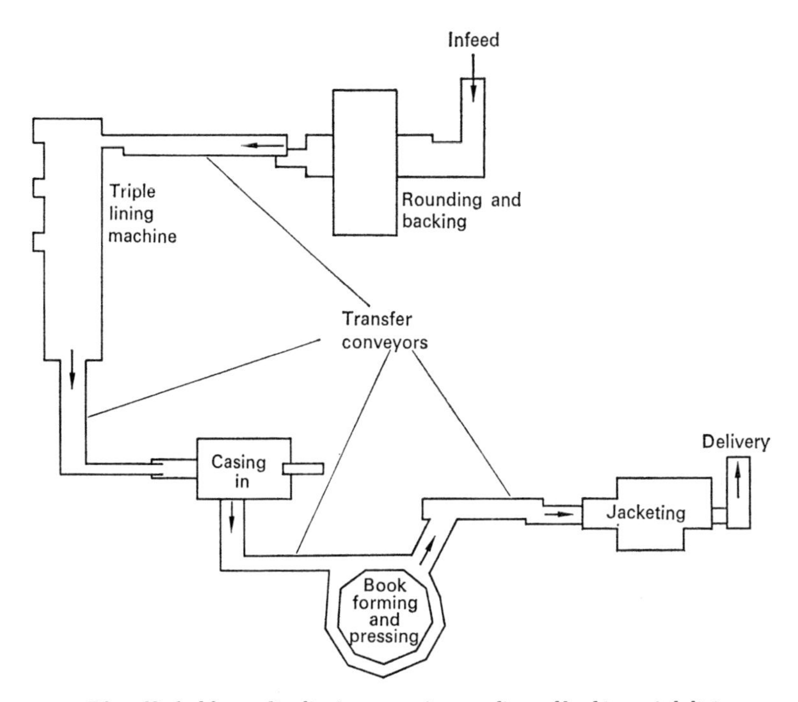

12.23.1 *Plan of linked forwarding line incorporating rounding and backing, triple lining, casing-in, book forming and pressing, jacketing.*

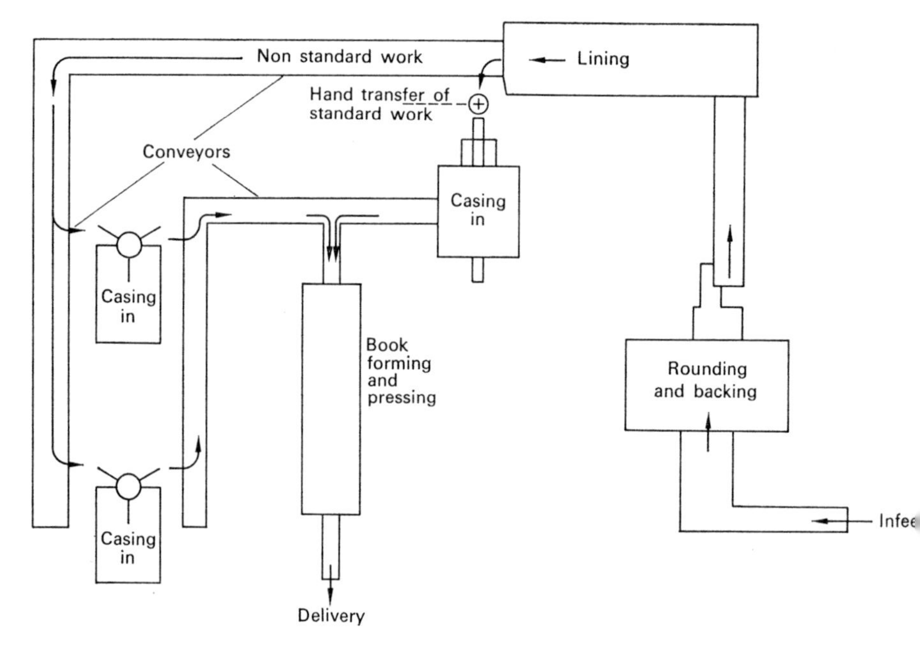

12.23.2 *A layout designed to accommodate both standard and large or small work outside the range of the continuous casing-in machine.*

job B. These factors and others that complicate the make-up have precluded successful automation in all except very long runs of uncomplicated work, *eg* pocket books and book club novels of standard dimensions.

To accommodate work of various thickness it is necessary to average out the jobs of different make-up and so balance production. If the equipment in a process sequence has an optimum production capacity of 20-section books, then jobs should be fed into production so that an average taken at any time will come close to this figure. If three jobs 1, 5 000 books of 12 sections; 2, 10 000 books of 21 sections; and 3, 1 000 books of 42 sections were processed through the plant in that order an average book value of 19·5 sections is obtained and the forwarding line would have sufficient work while the thick job (3) books were being sewn.

Variants

Whilst books from 100 × 65 × 6mm up to 325 × 265 × 75mm can be processed on linked forwarding lines, very large books are usually the subject of special arrangements that allow for flexibility and the extra care that large books should have. A typical arrangement for forwarding this work might be 1, glue and calico line on hand-fed machine; 2, three-knife trim and edges decorated; 3, first and second lining and headbanding on machine or bench as govened by dimensions 4, casing-in on three arm machine with 5, hand-fed book forming and pressing machine or built up in pneumatic press.

Diary production, while being in effect a case binding, is again controlled by size considerations. Almost all machinery produced for standard casebook work has minimum sizes too large for pocket diaries and large manufacturers have developed or modified machines to their own requirements.

Yapp bibles have a soft case with covers overhanging the edges and it is the production of this unit which separates bibles from the production of other publishers' work. The book block is made up and forwarded in the usual way although great care is taken to ensure high standards of craftsmanship; small pinhead joints are used and books usually have headbands and silk markers. Leather covers are cut on a 'clicking' press using a die of appropriate size and the edges are thinned on the paring machine. The prepared covers are now marked on the inside with a frame block on a blocking machine; this represents the completed size of the cover. When the cover stiffeners have been glued to the leather, the marked frame forms the creasing point for the glued turn-in so that all covers are guaranteed of identical size. When dry a further frame is blind impressed into the turn-in and this forms the folding point of the cover when in position on the book. After suitable cover decoration the spine of the book is glued to the cover (fast back) and the endpapers put down. Many variations exist including hollow backs and leather linings on more expensive versions.

13. Dielectric heating

One of the most interesting developments in recent years has been the introduction of dielectric heating into various aspects of print finishing processes. Dielectric heating is defined as the heating effect in certain non-conductive or insulator materials when they are subjected to or are adjacent to high frequency alternating voltage.

If a high-frequency voltage is passed between two plates separated only by air, a rapidly alternating stress is set up. In this condition no power will be absorbed from the system even though quite high voltages are used. If a solid material of poor electrical conductivity is placed between the plates, energy will be absorbed by the system and appear as heat in the material (fig. 13.1).

Materials that will heat in the high-frequency field are those which contain polar molecules. Although the individual molecule has an overall neutral electrical charge, the positive and negative charges within are displaced and

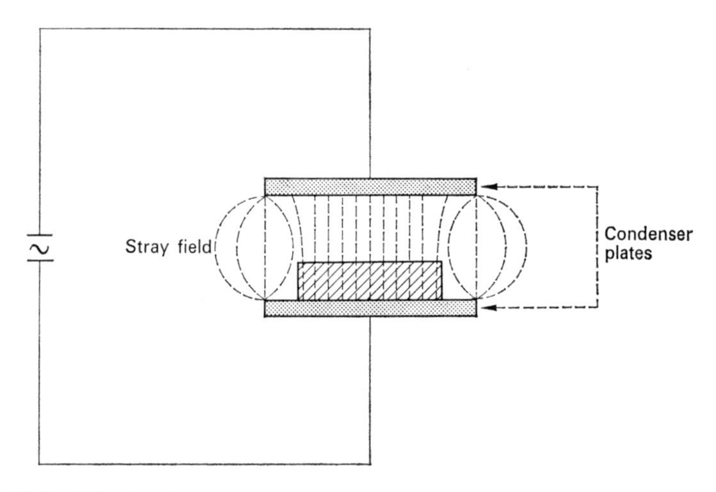

13.1 *Dielectric heating circuit.*

when subjected to a high voltage field will tend to twist or orientate to conform with the field. Each time the voltage is applied a stress is set up as the molecule tries to align itself and as the voltage is reversed at very many thousand times per second the stresses are likewise reversed. This induces what amounts to friction between the molecules and heat is generated.

Materials containing few dipoles and those in which the dipole moments are equal and opposite will not react to the field and will therefore be difficult or impossible to heat. These include polyethelene, polystyrene, silicone materials, electrical grades of ceramic, glass and many dry fibrous materials. Those that are easily heated include polyvinyl chloride in various forms, polyester films, acetate and many substances that contain moisture. In the latter case it is the water that responds to the field and often, when the moisture has been driven off, the material will start to cool whilst still in the field. This makes the drying of certain substances possible with little or no damage to the material itself.

Two main applications of this process are to be found in the print finishing area; high frequency plastic welding and the drying of various aqueous liquids.

Plastic-sheet welding

This application of the principles requires that two or more layers of polyvinyl chloride are clamped between two metal bars which act as the electrodes or condenser plates of the dielectric heating system. The high frequency field applied between the two bars creates the molecular friction previously described and heat is generated in the material. Because the shortest path in the field is between the two bars almost all the heating will occur in exactly the position required, *ie* in the shape of the electrode used. The heat will continue to be generated until the PVC is sufficiently softened for it to flow and under pressure generated by the press the layers fuse together in a weld. The bottom electrode may be in the shape of a flat machined work table covered by a 'barrier' material and this is usually the case when welding flat articles such as loose leaf covers and fancy stationery goods.

Because the outside of the top layer of PVC is in contact with the electrode and the outside of the bottom layer is in contact with the barrier material, both of which are cool, sufficient heat to soften the plastic is generated only on the inside surfaces or interface. If this were not so, the plastic would, of course, become a sticky mess and adhere to both tool and work table (fig. 13.2).

Electrode manufacture

The electrode die or tool may be manufactured from brass strip (modified printers' brass rule) or can be a shape machined from a solid unit. For most

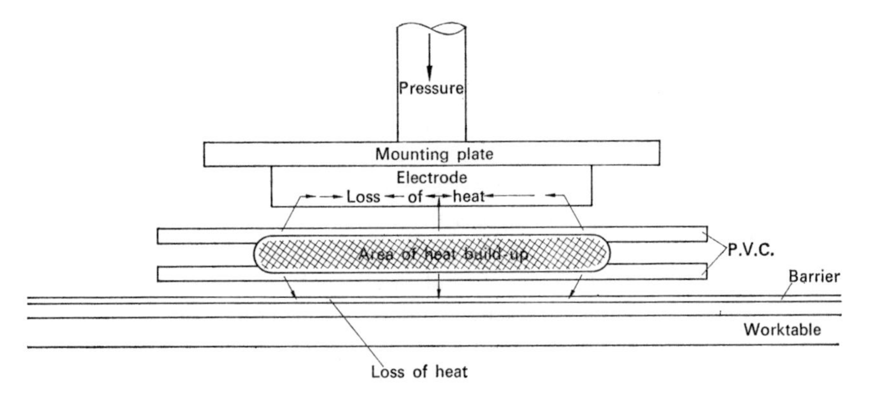

13.2 *High-frequency welding of polyvinyl chloride (all surfaces are in contact under spring or pneumatic pressure).*

loose leaf and bookbinding applications the former method is adopted, the rectangular shapes being relatively easy to construct in the workshop using simple metalwork tools. The brass rule used may have one of the profiles illustrated (fig. 13.3) and a face patterned to suit the particular requirements of the job. It will be noted that two dimensions are used: 23·3mm for the

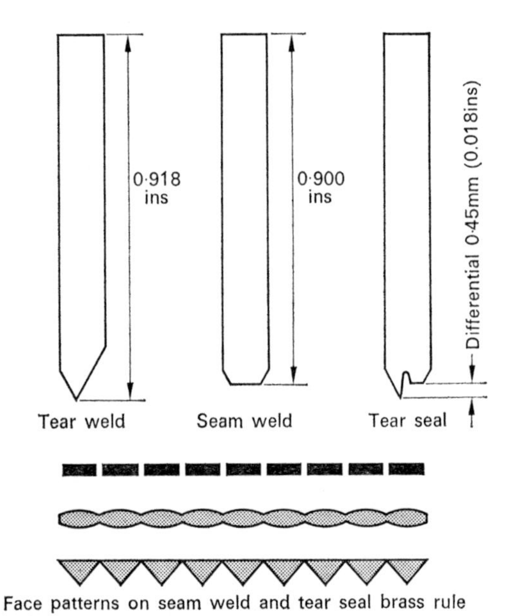

13.3 *Cross-section and face pattern of welding rule.*

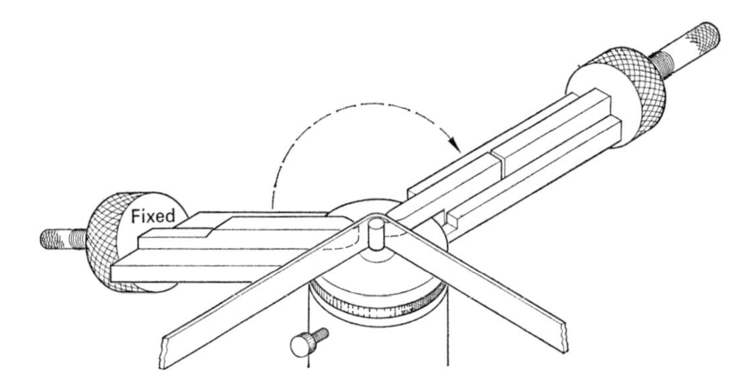

13.4.1 *Tool for bending brass rule.*

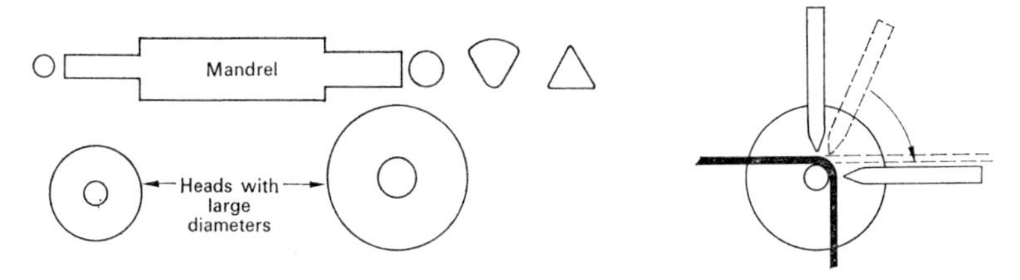

13.4.2 *Mandrel insert for bending tool.*

tear-seal and 22·8mm for the seam weld. It has been found that, theoretically, the difference between the weld and tear surfaces should be half the sum of the calipers of the materials used. For instance, an ordinary loose-leaf binder with inner and outer covers of 0·4mm material should be welded with an electrode that has a differential of 0·4mm, *ie* $\frac{2 \times 0{\cdot}4}{2}$mm. A similar product having pockets of 0·25mm material should be welded with an electrode having a differential of 0·5mm, *ie* $\frac{0{\cdot}4 \times 0{\cdot}4 \times 0{\cdot}25}{2}$mm. To prevent the need for a wide range of welding rule having differentials to suit each occasion a standard 0·4 or 0·45mm is adopted depending upon the class of work most likely to be produced.

The brass rule is cut, shaped (fig. 13.4) and fixed to a flat light gauge baseplate with any joints well soldered or brazed to maintain electrical contact (fig. 13.5). The outside shape is constructed from tear-seal brass whilst the hinges of the cover are made from seam weld material. Great care is exercised during the manufacture of the tool which must be to size and shape required and be in plane on the working surface. High points in the working surface tend to draw down extra power when working causing the possibility of excessive heating. It will also hold off pressure from other areas causing failure to tear cleanly. Many users pack the inside areas of the finished tool with foam material of sufficient thickness for it to protrude beyond the working face by about 3mm. This is intended to apply pressure to the layers of plastic thus squeezing out the air before it can be trapped inside the tool shape and become a permanent part of the product.

When electrodes for book and loose-leaf covers are made the hinge rule may butt on to the peripheral rule or stop short by 3mm or so. If the hinge weld joins the outside weld a potential weak point may be created; this is where many ring binders start to break down and a small gap left may increase the strength at this point. Similarly some manufacturers prefer a number of narrow welds along the hinge claiming increased strength and flexibility against the single weld.

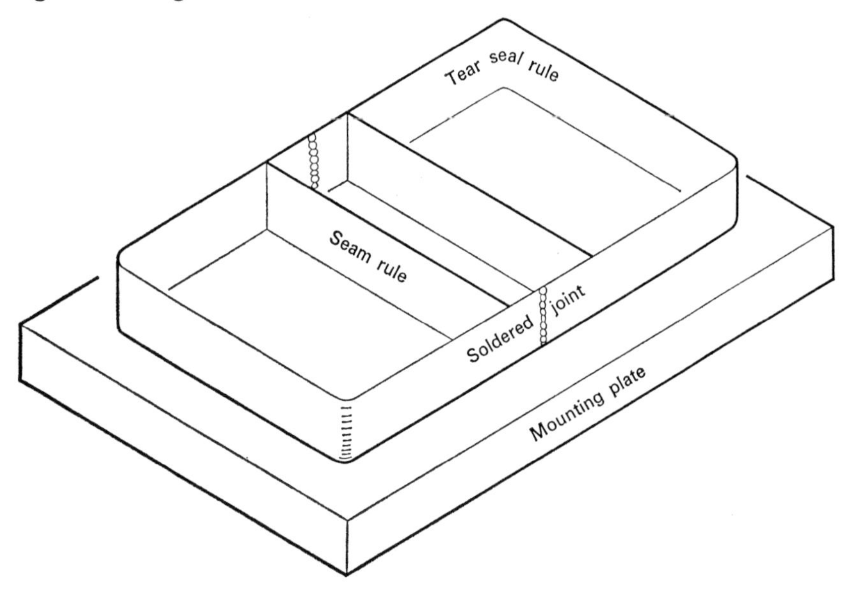

13.5 *Simple book cover electrode.*

When complete the electrode is mounted on to the press platen in such a way that it is truly in plane with the work table and a number of levelling screws are fitted for this purpose. The depth and weight of impression is then set for the caliper of PVC to be used. Pressure may be supplied by foot-

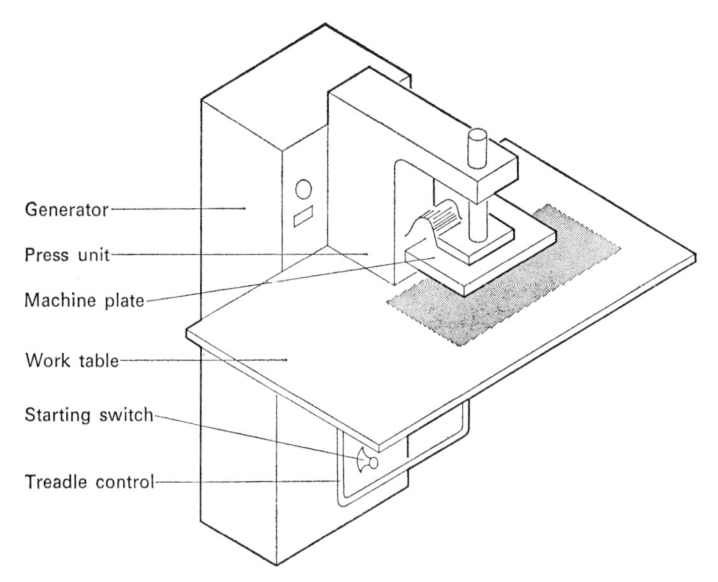

13.6 *Typical small high-frequency welding machine suitable for miscellaneous production.*

operated spring loading or by pneumatic power according to size of the press and productivity required (fig. 13.6).

The work table of the welding machine will normally be covered with a substance known as a 'barrier', the purpose of which is to prevent the blunt cutting edge around the periphery of the electrode from earthing when the welding field is in operation. Pahn board, varnished cloth, siliconised rubber, fibreglass and polyester sheet are all employed; material with a marked grain should be avoided as this will be taken up by the PVC while it is soft and show on the inside of the cover at the welded seams. Because barriers tend to act as thermal insulators they also help to increase the weld length per kilowatt available from the machine. As the surface of the barrier wears, fractures occur allowing the high frequency field to run to earth and create an arc at that particular point. This is both frightening to the operator and dangerous to the electrode which may be seriously damaged by the intense heat generated.

When the tool has been correctly mounted the power setting is adjusted at low value and the preset timing device set at high value. During a series of successive welds these two values are gradually brought closer together, *ie* by increasing power and decreasing the weld time until, from the visual appearance of the job, it can be said that the correct power is being applied for the correct length of time. It is usual for the press setting time to be adjusted so that pressure on the job is maintained briefly after the current is switched off. This ensures consolidation of the weld before removal from the press. Insufficient power will be observed as a lack of fusion of the constituent parts, failure of the waste to tear cleanly and lack of definition in the weld face pattern. Excessive power may cause the safety overload trip device to operate, though it is possible to set up the machine so that the PVC is seriously overcooked, without the trip operating. Overheating shows as exceptionally soft edges to the job as it comes from the press, very thin welds and excessive definition of patterns. If left on this setting the hinges of loose-leaf folders may be so thin that strength is lost and volatile plasticisers may be driven off and lead to early age hardening and fracture at the hinge.

A form of electrode useage that has some application in the production of very short runs and the making of samples is described as 'bar' welding. A miscellaneous machine has an attachment which allows a strip of welding rule to be quickly inserted and removed. By changing the length and type of rule used a straight-sided article can be built up one weld at a time.

Machine design

The two units that comprise the complete machine are the generator coupled to a suitable work press. Several alternative arrangements of the two basic units are possible and the generator may be placed at the side, behind, below

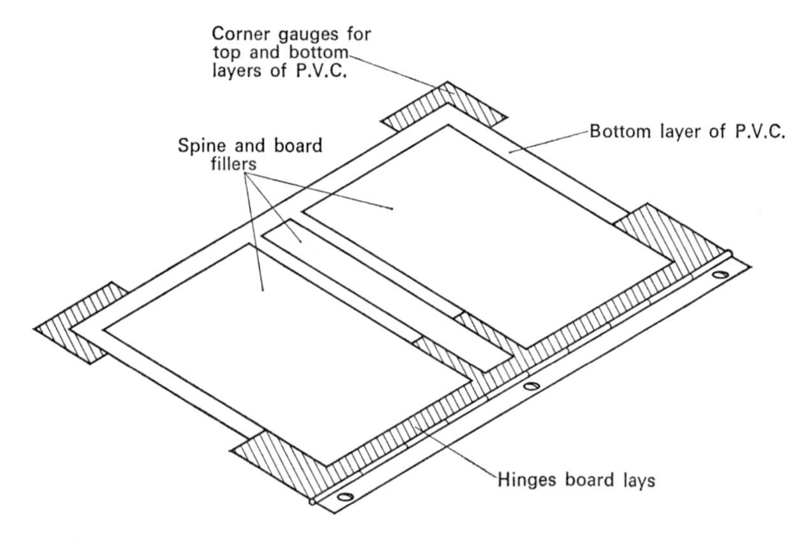

13.7 *Typical lay up gauges for simple three-piece case.*

or even completely separate from the press. Economy of space and the fitting of various feeding and handling attachments are among the chief considerations affecting the location of the two parts.

Small presses up to 3kW rating may be treadle operated but above this the platen area used calls for power operation by pneumatic or hydraulic rams. The treadle-operated versions are suited to the production of small runs and miscellaneous work. Typical ratings and related dimensions are given in the accompanying table.

Table 3 Summary details of typical low power HF welding machines

Rating	*Platen area*	*Approximate weld area*	*Equivalent linear dimension (When rule is 3mm thick)*	*Working frequency between*	*Pressure on platen between*
kW	mm	mm²	mm	mHz	N
1	570 × 300	250	80	70/73	264/880
1¼	570 × 300	450	150	70/73	264/880
2	610 × 430	660	220	48/58	880/3900
3	910 × 470	970	325	35/37	980/5900

Set up of the machine includes the provision of suitable lays and for short-run work these may be simple gauges built from binder's board and attached to the barrier with double-sided adhesive vinyl tape. For longer runs and those requiring greater accuracy in laying on, metal-hinged gauges are used (fig. 13.7). Simple constructions and particularly those not requiring the insertion of boards may be semi-automatically fed by a transfer mechanism. The reel stand accommodates up to three reels and by the movement of a handle predetermined lengths of the material is fed into the welding position of the machine.

Although the welding cycle may be as little as three seconds the assembly of some jobs may take 15 seconds or more and to make better use of the press time is the purpose of improved feeding mechanisms. The double shuttle system utilises two assembly positions one being under the press head whilst the other is being assembled. Another device with the same objective is the round table (fig. 13.8), which has three loading positions and either semi-automatic or automatic action with up to ten operations per minute.

Larger presses are described as pillar or bridge design and machines with ratings from 4–7kW are available in a variety of versions having double shuttle, rotary table, reel feeds and card inserting facilities.

Table 4 Summary details of typical higher capacity machines suitable for book covers and similar articles

Rating	*Lower platen area*	*Working frequency*	*Method of press operation*	*Power available to press*
kW	mm	mHz		kN
4 and 6	600 × 400		Hydraulic or pneumatic upstroke	4·9/6·9
7	650 × 450	50	Pneumatic downstroke	12·5

Machines with even higher kilowatt ratings are sometimes installed to cope with very long runs producing the article two or more at a time, but the complications of board feeding on these machines often slows the running time considerably (fig. 13.9).

Applications

Initially plastic welding was applied to the production of simple loose-leaf cases with board or metal hollows to which was affixed the ring metal, and developing from this came versions with gussets or pocket inside and label holders on the outside. Currently a very wide range of stationery articles are manufactured including clear wallets and folders, covers to be used in conjunction with mechanical and other forms of binding, zip cases, three-dimensional presentation boxes etc, and of course these often overlap into the fancy goods trade of purses, personal wallets, comb cases etc. Two particular applications which are quite difficult tasks in traditional materials are the provision of a window in the front board of a binder and the shaping of a

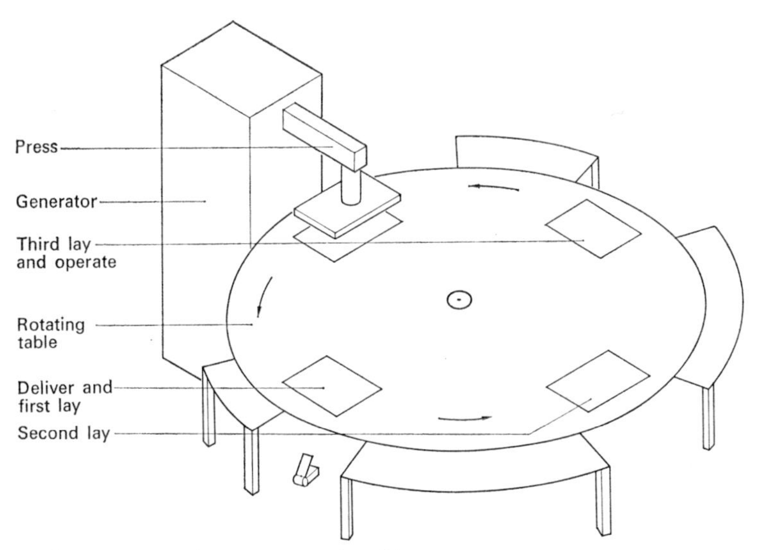

13.8 *Automatic or semi-automatic rotating table.*

strap closing device on products such as multi-year diaries, personal memo and account books etc. Both of these become relatively easy assembly problems although the manufacture of the electrode is rather complicated.

The decoration of plastic items

As with the assembly of any other mass-produced article in the print finishing field, decoration of the covers may be completed before or after the assembly

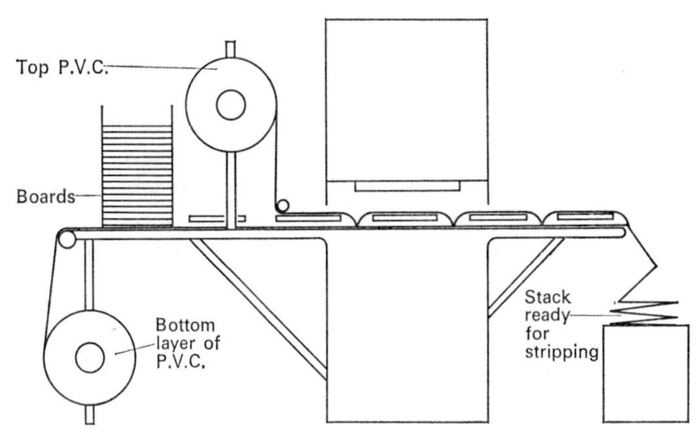

13.9 *Side elevation of automatic reel-fed press.*

is welded into its final shape. For purposes of accuracy it is frequently better to wait until the units are assembled before decorating, but this then becomes a relatively slow hand-fed operation. Where a cheaper product is involved extreme accuracy may be abandoned in favour of the ability to feed the sheet of PVC into printing machines at high speeds.

Most printing on PVC is done by screen process, usually before welding. Suitable inks are needed to ensure a good key on to the surface and metallic inks are very popular for this purpose. Metallic and carbon-based inks should not be carried across the weld area as these are conductive and will cause arcing during the welding cycle. The density of ink applied by screen printing gives good coverage of the rather strong colours and marked grain effects usually chosen for book covers. Letterpress, offset and gravure are all used for this purpose, the latter being particularly suited to long run overprinting in many colours.

Blocking

Traditional foil-blocking methods work very well on PVC providing no attempt is made to bottom the grain of the cover material. Because the material is very elastic, pressure tends to flatten the grain which then reasserts itself on release. The normal methods of applying heat do not work here either as the temperature necessary to soften the surface will overcook the foil size causing it to blister and bubble around the edges of the impression. The only certain method of flattening the grain is to use a light high frequency charge and sometimes an engraved block is incorporated into the welding electrode so that a trade mark is impressed into the product at the same time as it is made. When a metallic or coloured foil finish is to be placed on top of the blind impression two runs are needed and it is sometimes very useful to do this without moving the article being blocked. This can be accomplished by a welding-machine blocking attachment that carries the foil on a sliding head, the blind impression being completed before the foil feeding attachment is moved forwards so that the foil covers the block for the second impression. This attachment may have an electrically heated plate which can also pass a high frequency charge. This makes possible either normal 'hot' blocking, blocking by the so called 'cold' method using dielectric heating only, or a combination of both.

Appliqué

This is a method of decoration that utilises thin layers of plastic which are welded and cut *in situ*. The engraved brass block, which for this purpose has a fine cutting edge engraved into the edge of the image, is mounted on the heater block of the attachment mentioned previously. A layer of coloured plastic of between 0·1 and 0·2mm thick is laid on to the article and the impression made. As the heater box and the block are slightly warm the PVC is softened and the high frequency field operating through the engraved brass block welds the periphery of the image to the parent article. The blunt cutting edge around the image cuts through the thin PVC allowing the waste to be torn away as a subsequent operation. With fine detail work the removal of the waste may be a difficult task unless tweezers and needle points are used.

When setting the press for this work very accurate depth and pressure setting allied to a carefully controlled high frequency time cycle is needed. Insufficient depth or power will result in difficulty of removing waste, whilst oversetting may injure the cover to which the decoration is being attached. Multi-coloured appliqué work is achieved in one stroke by having the engraved brass suitably stepped so that, in use, the cutting edges are at two or more levels. In this way the single weld will secure one layer to the cover and the second layer to the first. By careful design and block cutting up to three layers can be successfully welded in this way.

Appliqué tends to be an expensive operation, completed at low speeds with higher than average wastage, and this renders it unsuitable for any but the most highly priced articles. One manufacturer offers a specially cut brass appliqué type in a small range of sizes and suitable for short run miscellaneous work.

Other decorative methods

The difficulty of printing close register multi-colour designs on to PVC has led one company to try another approach. The design, say a trademark, is printed on to a release paper which is laid into position on the article to be marked. A simple flat plate shaped to suit the design is mounted on the press and a short charge applied. This welds the multi-coloured design to the article, the release paper being peeled off and discarded. Although not a direct competitor to blocking in either price or brilliance it seriously challenges screen process printing for its attractiveness and ability to present several colours in good register. It is considerably cheaper than appliqué and the presentation of much finer detail is possible with this technique.

A very simple method of illustrating the covers of loose leaf binders is to use clear PVC and line the board fillers with the illustrations required. This, of course, involves the extra operation of board lining before make-up but when customers insist upon, say, a halftone of a machine in the cover of a catalogue, this is the only reasonable solution.

Other applications of dielectric heating

Plastic-sheet welding apart, dielectric heating is being adopted only very slowly in other fields of print finishing. This is probably because of the high capital cost involved and a basic lack of appreciation of the qualities of the process. It is particularly suited to the fast drying of aqueous adhesives, *eg* animal, starch and PVA formulations, the water content being driven off as steam.

When drying the PVA primer on a two shot adhesive binding machine the wet book passes directly through a high frequency field whilst still in the clamp and this leaves the adhesive dry but hot. This is quite satisfactory as the hot melt application follows very shortly afterwards and the spine can be considered in a pre-heated condition and ready for that application.

The curing of books that are adhesive bound with PVA only is a different proposition as the high-frequency field has to locate the glue line between the book and its cover. In any case this is best completed whilst there is pressure applied to the spine and this is the subject of a special piece of equipment powered by a 25kW generator and includes a cooling potential.

Another solution to this problem is to pass the newly wrappered books along a table on a conveyor belt. Beneath the table electrodes have a high-frequency field passing between them and are so positioned that the book spines are in the stray field.

Drying off the starch/PVA adhesive used in casing-in books is another area where the application of dielectric heating is progressing. The cased in volume is fed in random sequence between two conveyor belts mounted vertically so that the book is under light pressure as it progresses. In the centre of the run extra springs exert force on the conveyors while the book proceeds through the high-frequency field. The molecules are agitated creating heat within the adhesive which then dries whilst the book is under pressure and allowed to set and cool on the run out. One machine in this field uses a 3kW generator and processes work at 25m per minute on a belt 230mm wide (fig. 13.10).

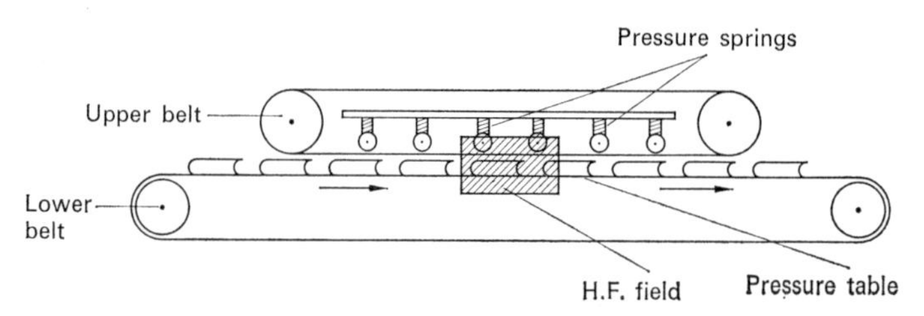

13.10 *Dielectric book pressing machine.*

The simplest of all these devices is a conveyor belt that passes sheets of cover material coated with aqueous adhesive through a high frequency field. This work is usually dried in a long infra red tunnel and the use of high frequency reduces the machine length needed as well as being more economical in running. The power needed to dry off the sheets is only used when work is actually in the field.

Many other applications appear to be ripe for development and may in future contribute to the production of true in-line systems. These include drying of adhesive on first gluing and calico lining machines; curing the adhesive on triple lining machines so that books are ejected dry and in good shape; perhaps as the drying mechanism on a case-making–blocking link-up; drying certain varnishes and carton sealing adhesives.

In the associated production fields applications appear in microwave drying on web offset and wallpaper-printing machines, in the production of paper and fibreboard where it is sometimes used to extract the last vestiges of moisture, if this is desirable; and it may also be used in the moulding of rigid PVC matrices for duplicate plate-making.

Operational safety

The operative can receive a small but irritating burn if the electrode is touched during the welding cycle. This is a distinct possibility on the miscellaneous hand-fed/foot-operated models and operatives have to adopt a feeding and operational rhythm which inhibits such an event. All power-operated varieties can be arranged to be started by the double-button system which makes accidental touching of the live electrode much more unlikely. Double pillar and bridge type presses with shuttle attachments usually combine interlocked operative guards as part of the standard equipment.

The principal production hazard on machines is that of ‘flashing’, ‘sparking’ or arcing and if this occurs damage to the electrode, PVC material, barrier (and operative’s nerves) may be caused. Reasons for sparking may be sought among the following:

1 Adjustment of the power setting so that peak voltage between the electrodes leads to a voltage breakdown for that particular quality and thickness of material which in turn leads to a spark passing through the material at this point.
2 Defective material, *eg* conductive impurities in the PVC, small pinholes, badly dispersed pigments, etc.
3 Presence of moisture or electrically conducting printing inks on the PVC, *eg* those containing carbon black and metallic dusts. It should be

noted that carbon lead pencil marks will act in the same way as carbon black printing ink and pencil marks on the PVC should always be avoided.

4 Tracking caused by the electrode projecting beyond the edge of the PVC.

5 Imperfect barrier, *eg* fractures caused by constant use, holes or carbon charring caused by previous spark.

6 Excessively sharp edges and corners on the electrode or the electrode in bad plane alignment with the work material.

7 Freak high voltage caused by faults in the generator, fluctuations in the mains supply, etc.

14. Adhesives and adhesion

If without the aid of mechanical means two surfaces are held together in close contact by placing a third substance between them, they can be said to be adhering; the force obtained arising from two quite different effects.

Mechanical or physical adhesion in the paper industries is achieved by the adhesive flowing into the structure of the adherand or around the surface irregularities and producing a keying effect, but when the surfaces to be stuck are impervious and smooth, adhesion will rest upon the molecules of the adhesive having strong attraction for the adherand. This is usually called 'specific' adhesion. Sometimes a combination of the two principles may be needed to achieve adhesion between two widely different surfaces.

The strength of bond between two surfaces must be such as to resist the forces that are trying to separate them. The bonds obtained in paper products are often greater in strength than the materials being held and clearly this means that the adhesive is adequate. If under stress the adherands separate leaving some adhesive on each half, then the internal strength or cohesion of the adhesive is inadequate for the task.

Factors governing the choice of adhesive are:

1 The type of surfaces to be joined.
2 The time needed to assemble the parts of the job before the adhesive sets off; sometimes called the 'open time'.
3 The degree of tack needed. 'Initial tack' is the ability of an adhesive to resist separation after two surfaces have been joined but before the adhesive has set or is dry. This is very important in the high-speed assembly of bookcovers and rigid boxes.
4 Method of application. The use of brush, roller, glue wheel, etc, and any special problem involved, *eg* 'frothing' and 'flying'. Flying is when the adhesive is thrown off the applicator roller because of the speed of rotation.
5 Special requirements of the adhesive relating to flexibility, resistance to mould and insect attack etc.

6 The degree of pressure or heat that may be used to assist the bond whilst the adhesive is setting.
7 The length of holding pressure (pressure time), if any, and how long before the adhesive must be fully dried out (drying time).
8 The economics of the adhesive.

Adhesives are variously described as glue, paste, gum, hot melt or cement and this usually refers to either its condition at room temperature or the manner in which the adhesive sets off. During their application it is generally necessary for adhesives to pass through a fluid or near fluid stage and then to set into a solid condition.

A convenient method of classifying adhesives used in finishing operations is by means of the main component used in its manufacture.

Animal glues

Manufactured from residual animal products, this is the traditional adhesive that has been in general use since the dawn of manufacturing processes.

Waste bone is cleaned and crushed into small pieces. After a maceration period to remove unwanted blood and mucin the bone is sealed into a digestor tank and subjected to pressurised steam. In this way the collagen is released from the bone and converted to a gelatinous liquid and by repeating the process many times glue liquors of various strengths are obtained. The blending of these produces the various qualities offered for sale. Hides and skins may also be used, rabbit pelts producing a particularly strong adhesive.

Formulations of animal adhesives includes substances to preserve, bleach and plasticise and in this way it is modified to suit various purposes. Flexibility and non-warp characteristics are obtained by the addition of hygroscopic substances such as sorbital, glycerol and glucose. Colour may be lightened from the natural tan to a near white by the addition of sulphur dioxide or hydrogen peroxide and pigments added as required.

Putrefaction of animal adhesive is caused by air-carried bacteria and moulds; preservatives are added to prevent the action of these harmful agents. Other additives include inert fillers to increase the solids content, anti-foam substances, perfumes and deodorants. With the formulation complete the liquid is now evaporated off to leave the required solids content and converted into one of the forms needed by the customer. These are:

Prepared cake

Large blocks (5–10 kilos) wrapped in film and in a condition suitable for immediate addition to the pot or machine tank.

Pearl or bead

Small round pieces of glue 2–4mm in diameter. This variety should usually be soaked and allowed to absorb water for a short while before cooking otherwise 'stringing' will result.

Powder

A biscuit-coloured powder which is usually mixed with water before heating. Powders to produce cold glue are also available.

Animal adhesives are best used in water-jacketed boilers and kept at a temperature below 42°C. Excessive heating will seriously reduce the strength and glue kept at 60°C for ten hours may lose up to 30 per cent of its original strength.

Speed of setting

A glue chills at the moment it reaches gelling point and then sets by losing its moisture. The speed at which this occurs is governed by:

1 The quality of the glue.
2 Percentage of solids present – with more water, the glue is slower to set.
3 Temperature of glue when applied.
4 Temperature of substrate. Hot glue applied to a cold substrate produces quick chilling and minimum penetration.
5 Atmosphere of environment. A cold dry atmosphere produces quick chilling (excessive draughts produce the same result). A warm moist atmosphere increases the time required to chill and to dry.
6 Thickness of coating. Thick coatings hold heat longer and are therefore slower to chill.

A good animal adhesive will conform to the factors of choice previously mentioned and also be of good colour, have an acceptable odour, perform successfully and have a reasonable shelf life. Modern packaging ensures that packs of animal glue remain in good condition providing the store is reasonably dry and cool.

Vegetable adhesives

Adhesives derived from vegetable sources are described as 'paste' or 'gum'. Starch from maize, potato, tapioca and similar sources are cooked in water. This allows the heavy molecules to intertwine and produce the short slow moving, non-stringing substance recognized as paste. Simple pastes are usually slightly acidic in character taking their pH from the raw material from which they are made; but as many paste formulations contain borax and other alkaline substances these take the adhesive into the high pH ranges. Low pH pastes tend to hydrolise and thin out upon ageing.

Starch pastes contain a high percentage of water and, although this makes for maximum penetration into the adherands, excessive expansion and slow drying times are both experienced as well as a lack of flexibility.

Gums are usually liquids of moderate viscosity which are useful mainly for their remoistenable characteristics. Natural resins in the form of exudation of acacia trees (gum arabic), dextrine and modified starches are the main ingredients. In some traditional gum-using areas of production, *eg* envelope manufacture, synthetic resin adhesives have been introduced as alternatives to the vegetable gums normally used. This is particularly true in the attaching of acetate and other films as windows in envelopes and often blends of synthetic and vegetable adhesives are found to be useful.

Used cold they should be odourless, tasteless and non-toxic; after remoistening the tack should be high.

Synthetics

Synthetic resin emulsions and 'hot melts' are the two main synthetic adhesives used in finishing operations.

Polyvinyl acetate (PVA) emulsion is more strictly referred to as an aqueous dispersion, the solids content being finely dispersed to produce a white substance of thick cream consistency. Emulsions dry by loss of moisture content and after several hours a thin application of the adhesive will appear as an opaque and flexible film having theremoplastic qualities. Formulations vary widely and initial tack, open time, solids content and other characteristics may all be readily adjusted by the manufacturer. A feature of these adhesives is that, unlike vegetable and animal types, they can be formulated to adhere to impervious surfaces. In print finishing this is a useful quality when plastics, *eg* polyvinyl chloride, polythene, polyolefin and polyester are being used in film laminating and in the bookbinding fields.

The PVA's offer no attraction to starch-feeding pests such as cockroaches and even under high conditions of relative humidity do not easily provide a growing media for mould and fungi. This makes these adhesives very suitable for situations where books, etc, are stored in tropical climates and are subject to these problems.

Although industry has adopted PVA adhesive in many areas of production, certain limiting factors have to be recognised. These are drying time and initial tack. Drying time is particularly important during in-line production and where these adhesives are used in the production of magazines by the adhesive process and allowed to dry naturally, a drying time of not less than two hours is usually required making high speed in-line production very difficult.

The initial tack qualities of PVA adhesives is being steadily improved but has not yet reached the point where it seriously competes with animal glue in this respect. As most lining and casemaking machinery relies upon the tack of the adhesive for its high speed of production PVA's are being adopted very slowly in these areas. But when they are used in this context the resulting adhesive film has high flexibility and will resist both animal and fungi attack.

Hot melts

Although this term, applied to finishing adhesives, refers to a range of synthetic substances, strictly speaking it is descriptive only of its physical condition and in no way indicates its chemical formulation, which can vary widely. This adhesive is a 100 per cent solids construction of polymers, resins and waxes. It requires heating to between 150 and 170°C to liquefy and sets off a few seconds after being removed from the heat source; *ie* the 'open time' may be from 2 to 15 seconds.

The first hot melts had a reputation for being rather toxic when in the liquid condition but manufacturers are now stressing that the present ranges are not toxic in any way. If the temperatures needed to maintain the liquid condition are exceeded by many degrees the waxes tend to burn off and for this reason fume-extraction units are frequently fitted above the glue pots of continuously running machines. Overheating of the adhesive tends to degrade it and when seriously overcooked it becomes unusable.

Hot melts may be supplied in pieces, billets or plugs and in continuous rope form. The ability to shape the adhesive at will makes possible many variations of glue applications including nozzle and gun versions that may be loaded with a small pellet or plug, or may be continuously fed with a rope adhesive from a coil. In many installations preheaters are used, the melted adhesive being pumped through heated pipes to the point of application.

Because hot melts are thermoplastic in character they can be used to pre-coat surfaces that are afterwards reactivated to produce the required tack. This facility is useful in situations where the two adherands are treated with hot melt and allowed to cool. Later the two surfaces are brought together under pressure and heated to complete the joint. The rapid acceptance of hot melt as an adhesive in finishing is due, almost exclusively, to the high speeds which are obtainable on manufacturing machinery; in some instances, after suitable modification speeds have doubled.

The main disadvantages of these adhesives, other than fuming, are high cost and the tendency to crack at low temperatures. Although this feature is not vital in book production in the United Kingdom (unless the book is being exported to very cold climates), there may be some need to consider this problem in relation to packaging, *eg* where a carton is to be stored under refrigerated conditions. A factor that has to be considered in relation to the use of hot melt is that of waste disposal. Trimmings from adhesive binding lines using this adhesive command a much lower price than is usual for white waste because the mills are at present unable to repulp this material successfully.

Use of adhesives in print finishing

Starch paste
Tipping on plates and endpapers; paper-mending and joining processes; attaching endpapers to cover; covering books with leather and some hard fabrics.

Gums
Some tipping machines; edge and sheet gumming for remoistening situations; envelope manufacture.

Animal glues, scotch
Assembly of hand bookbinding units, *eg* attaching boards and cover materials; showcard lining and covering; boxcovering.

Animal glues, flexible
First gluing of book spines; casemaking and lining of publishers' case books; modified versions for adhesive binding.

Polyvinyl acetate (PVA)
First gluing of book spines; hand binding operations; some first and second lining machines; adhesive binding of both wrappered and lined work; specialist casemaking especially diaries and yapps; foil laminating; boxcovering; carton gluing and sealing.

Hot melt, one shot
Adhesive binding of magazines and paperbacks by machine; carton gluing and sealing; adhesive binding/lining of some publishers' case work.

Hot melt, two shot
Adhesive binding on certain machines only.

Volatile adhesives
Foil laminating.

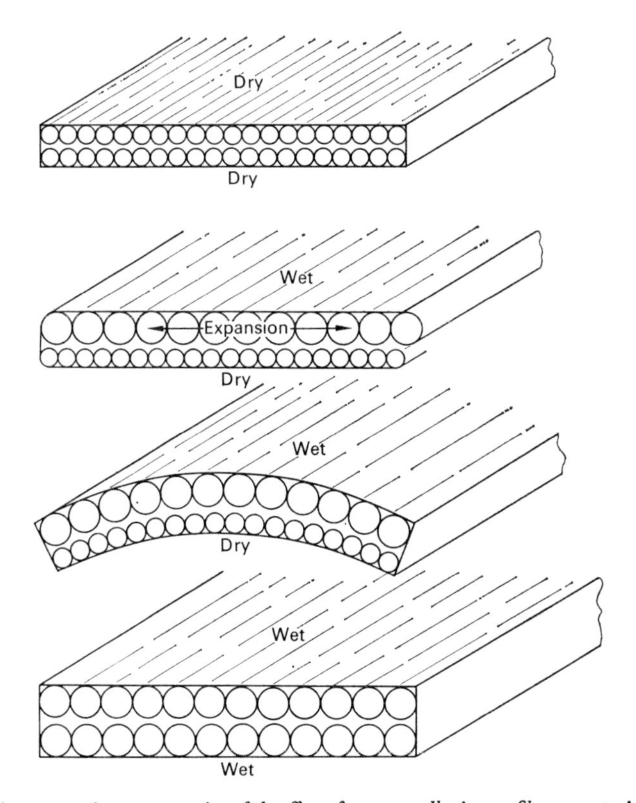

14.1 *Diagrammatic representation of the effect of aqueous adhesives on fibrous materials.*

Effect of wet adhesives on paper and board

Perhaps the most fundamental relationship in print finishing processes lie in the reaction of paper, board, cloth and leather to each other in the presence of aqueous adhesives. To appreciate fully these effects it is essential to have a good understanding of the construction of fibrous materials and the way the shape of these are modified by wetting and drying.

Paper, binding board and bookcloth are all based upon cellulose fibres which have a tubular shape. Given the opportunity, these will absorb water by capilliary action and as a result the tube will increase its diameter considerably, but gain only fractionally in length. Paper and board are composed of these fibres matted together in thin sheets with about 70 per cent of the fibres lying in a more or less parallel direction (grain or machine direction). As each fibre expands in diameter, the collective effect upon the matted fibres in sheet form is expansion across the machine direction or grain.

The application of an aqueous adhesive to the surface of a sheet of paper will cause the fibres on that surface to expand rapidly but, because there is a time lapse before the moisture has penetrated evenly through the stock, expansion will initially be uneven and a curl will be induced into the sheet (fig. 14.1). The material will not flatten until there is an even distribution of moisture and the time lag before this occurs will depend upon the wetness of the adhesive, the furnish of the stock and the environmental conditions pertaining at the time; the amount of expansion will also be controlled by the first two factors. As curly materials are difficult to handle, the practical bookbinder will often glue and lay out several articles that are to be assembled, allowing the short time lag necessary for the sheet to flatten; this also allows the adhesive partially to dry thus improving the initial tack on the surface. To slow down the drying of the adhesive without preventing it from flattening, two glued sheets are 'faced', *ie* the glued surfaces are placed face to face for a short while.

When the sheet so treated is attached to a piece of strawboard, the board reacts in much the same way as the paper, *ie* moisture from the paper wets the strawboard so that the surface expands and the whole board curls. As the board is very much thicker, the effect may be slower in showing and with very thick strawboards insufficient fibres will be wetted to have any noticeable effect upon the remainder.

After the two substrates are joined (the paper in extended form) moisture will be lost from the assembly at a rate and to a point depending upon the environmental conditions in the workshop at that time. As moisture evaporates from the paper it will attempt to contract back toward its original dimension, but as it is joined by the adhesive to the strawboard, which has expanded only marginally, the paper will pull the strawboard into a curl or warp (fig. 14.2). A complicating factor in the expansion and contraction of paper is the built-in stresses normally associated with web-made materials. During manufacture the speed of drying and the control exercised over it by the felts prevents paper from contracting as far as it might if left to dry uncontrolled. When wetted some of these in-built stresses are released and the stock may appear to contract further than it initially stretched.

This activity of paper and other fibrous materials expanding and contracting in the presence of moisture occurs in bookbinding, box covering, showcard manufacture, etc, and wherever possible a balance should be obtained by lining both sides of the strawboard. When choosing substances for lining, account should be taken of the nature of the furnish. Papers with short fibres tend to saturate more rapidly and therefore expand and contract quickly. Long-fibred stocks are much more stable in this respect and in certain

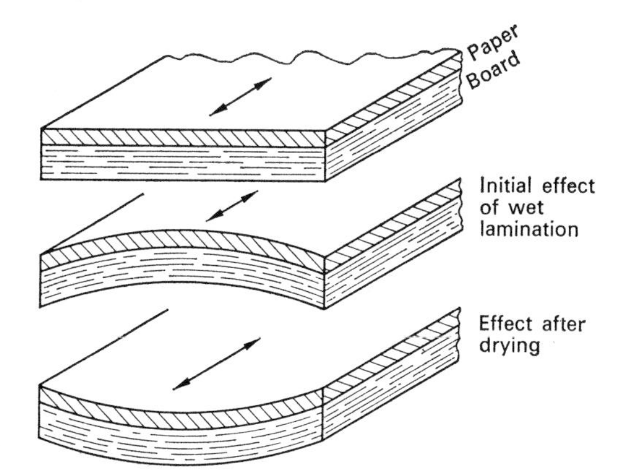

14.2 *Effect of wet paper on strawboard.*

instances it is possible to exercise considerable control over the ultimate shape of an assembly by balancing off the substrates in this way.

The woven threads of bookcloth fabrics respond in much the same way as the fibres of paper; the warp threads, *ie* those travelling the length of the roll, corresponding to the machine direction of paper.

Once accepting that glued paper will pull or warp a sheet of board when attached to it, it is necessary to decide how best to minimise the effect it may have on the assembly and construction of items produced. In bookbinding and showcard mounting it is usually undesirable to cross laminate the grain direction of the substrates as this will stress the centre board and produce considerable distortion.

In case books, text stock and endpapers are both subjected to wet adhesives when the endpapers are attached; and this occurs in a long narrow strip parallel to the spine. If the endpaper has long grain the expansion can occur without distortion but when short grained, only a narrow strip is attempting the expansion and cockling results in a bad attachment. This effect may also be seen on certain work when a book on cross-grain text is glued on the spine with thinned polyvinyl acetate showing a distinct cockling of the text pages which may never dry out.

A further factor is the 'pull' exercised by the endpaper on the cover of the volume. When parallel with the spine any ensuing curl will be across the width of the book (fig. 14.3) and this allows the long edge of the book board to stay in the shoulder housing. If cross-grain endpapers are used a contrary pull will develop so that the long edge of the board will lift out of the shoulder housing toward the centre of the book and the board opening will not function correctly (fig. 14.4).

All fibrous materials also fold and crease most easily along rather than across the fibres. Text and endpapers will open more readily and the book will lay flatter if the grain is parallel to the spine. Linings of books will mould and open more readily for the same reasons.

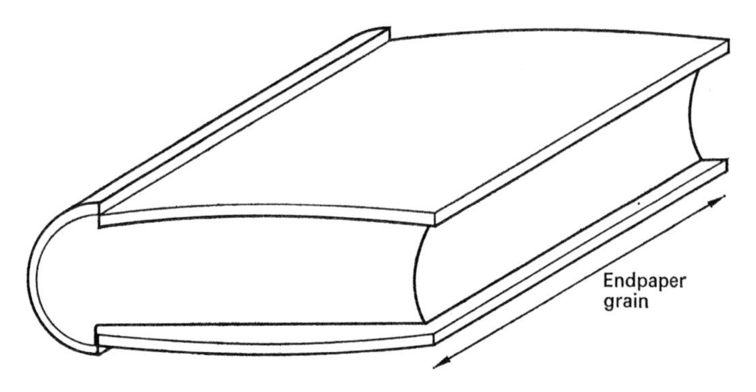

14.3 *Effect of long grain end-papers on book boards.*

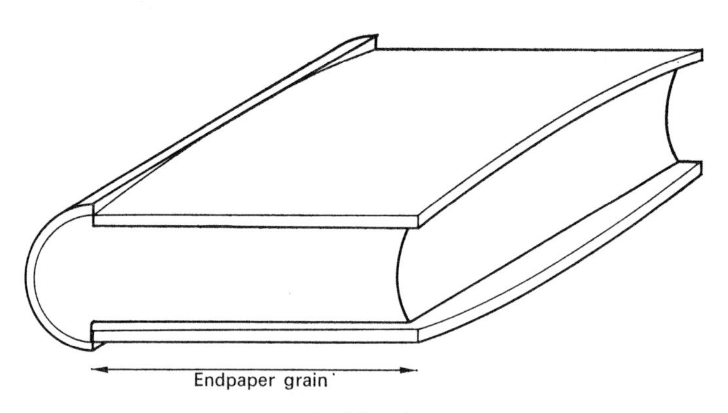

14.4 *Effect of short grain end-papers on book boards.*

Adhesive-bound books tend to be a special case; almost all paperbacks are printed on web-fed presses delivering sections with short-grain text. When these are reduced to single leaves and coated with aqueous adhesive, considerable distortion of the spine will often result. This can sometimes be seen on the finished job as a wave or cockle showing through the cover. The solution is the use of a non-aqueous adhesive.

Laminating

There are two aspects to laminating, which is the joining together of thin layers or laminars. One is the lamination of polythene, cellulose acetate, polyester, etc, to form clear materials for packaging, particularly for the food industry. The advantages to be gained by laminating are improvements in printability, gas and moisture barrier qualities, impact and burst strengths and many others. This form of lamination is a reel to reel operation carried out at high speed on very large machines and does not normally come into the print finishing field.

The lamination of printed paper to film can be either a reel- or sheet-fed process and may be applied to a wide variety of display material, boxes and cartons, record sleeves, book jackets, etc. Another important but less used area is that of document protection, *eg* archive repair, map lamination and book repair; an associated field is dry mounting for photographic and art work. All these may be the legitimate province of the print finisher.

Laminating films include cellulose acetate, orientated polypropylene (OPP), polyvinyl chloride (PVC) and polyester in calipers from 0·0125mm to 0·025mm with a variety of optical and mechanical qualities available within each type. The method of joining the two substrates together may be by the wet or dry process. The wet process requires that the film be coated with liquid adhesive of either the aqueous or solvent types. The latter are preferred, as this enables the machine length to be shorter but has the obvious disadvantages of inflammable solvents, toxicity and possible difficulty in evaporation of the solvent. Solvents used are toluol, ethyl-acetate, isopropyl-

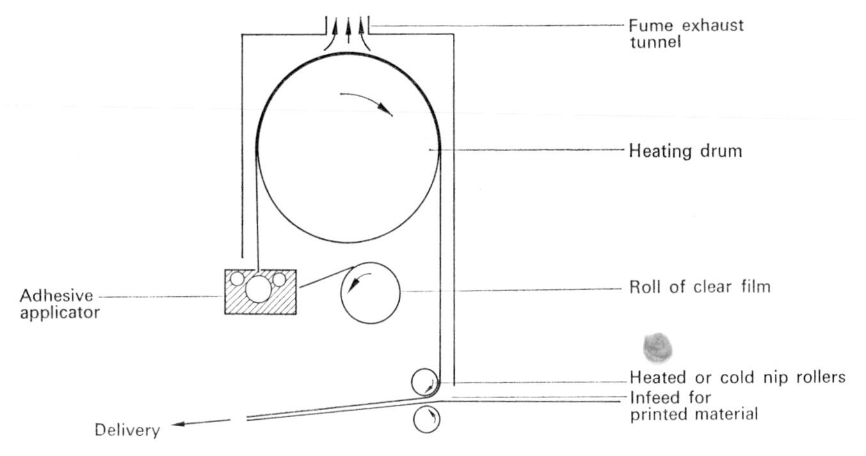

14.5 *Small laminating machine using heated roller to evaporate off the adhesive solvent.*

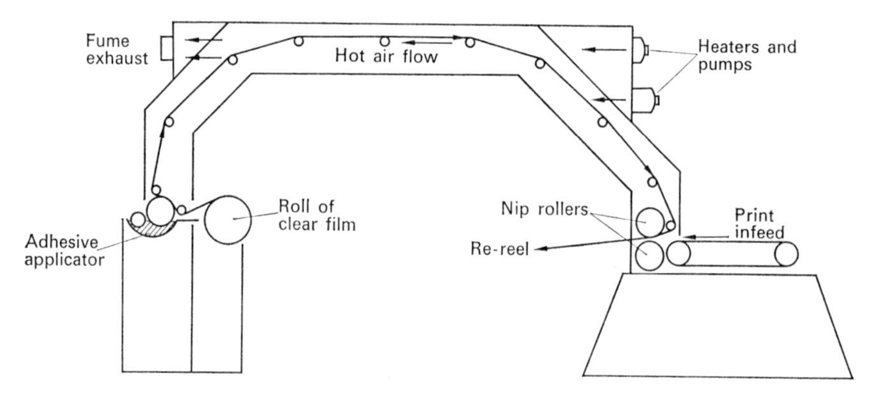

14.6 *Tunnel laminator using hot air to evaporate the adhesive solvent.*

acetate and trichlorethylene. Dry laminating implies that the film has been pre-coated and dried before being used and two types are available; one has a dry non-sticky film that is reactivated by heat and the other has a tacky layer protected by a removable release shet.

Machines for wet laminating have a number of common units; a continuous reel of film; an adhesive coating system; a drying tunnel or roller; hand-feed board or paper feeder for paper input; laminating nip rollers; cut-off table or re-reeling position (figs 14.5 and 14.6). The film is reeled off and passed through the roller adhesive applicator; to obtain optimum conditions the solvent adhesive may be circulated through the system by a pump and is filtered before being returned to the machine. The film web is then carried through the drying system of a tunnel or rollers and is subjected to direct or air-blown heat to evaporate off the solvent which is extracted and blown off into atmosphere; this leaves the film with an even coating of tacky adhesive.

The printed stock is now fed by hand or from a pile feeder, so as to be slightly overlapped and these make immediate contact with the tacky film in the roller nip; the rollers here may be hot or cold according to the job being worked. Almost immediately the work may be slit apart by hand or re-reeled for subsequent processing. The over- or underlap feed is necessary so that no area of the sticky film makes contact with the nip rollers, but this does mean that about 5 mm of the printed surface is not laminated on one edge and this has to be planned at the printing or design stage. The width of the film will always be narrower than the print for the same reasons and this implies that a trim is necessary if the film is to appear right up to the edges of the completed job. For certain jobs one or more strips of film may be laminated to cover only part of the printed surface.

A simple test of the success of lamination bond is to tear the layers apart, noting how the adhesive pulls off the ink and perhaps the paper fibres too. Generally speaking excess adhesive is uneconomic, may effect the optical qualities of the finish, and even show as an orange-peel effect. Patchiness may be caused by starvation of the adhesive and various bleeding and ink-discoloration effects may be due to residual solvents.

Care in choice of inks and stock is necessary when lamination is considered with consultation and planning before printing begins. It is a mistake to laminate as an afterthought as optical variations are brought about by refractive and reflective qualities of the film. Ink often appears to have greater density after lamination and best results are achieved by using quick-drying matt inks; several days drying and curing is desirable before lamination occurs. Faults in printing and stock surface are often emphasised by the lamination; anti-setoff sprays constitute a barrier to good adhesion and should be avoided or kept to a minimum.

Dry laminating is used more in miscellaneous work than in large volume production. The heated process uses film and tissue pre-coated one side with heat-sensitive adhesives. The layers are brought together in a press having a heated platen or roller system; the adhesive is reactivated and becomes tacky. Laminating to one or both sides of the sheet, mounting or combined laminating and mounting is within the range of the platen type press.

The pressure (or self) adhesive films have a release paper to protect the adhesive coating and this has to be peeled off before use. Although machines are used for this work many binderies make use of this material for the protection of charts, display material, bookcovers, etc on a miscellaneous basis, *eg* the laminating of book jackets that are afterwards used for library rebind cases. Care must be taken to bring the two surfaces together so that air is excluded otherwise bubbles and creases appear in the surface.

15. Adhesive binding systems

The extension of the simple methods of adhesive binding into the high-speed automatic systems used for the production of paperbacks and magazines is one of organisation and scale rather than of principle. A typical binding line for this class of work involves all the functions previously described but as speeds are much higher all standards are more critical. Units that are usually included in a single automatic line are gathering, adhesive binding/wrappering and trimming. For special purposes, mechanisms may be added to dry off the adhesive, insert loose material into the bound volume, count and stack, pack or wrap for mailing.

The first and probably the weakest part of the line is the gatherer which will have sufficient hoppers to cope with the particular jobs to be undertaken and may be of the arm, rotary or other types. The speed of this unit is usually controlled independently of the binder thus making it possible to gather thin books two up at half the operational speed of the binder. As with gathering in publishers' case binding many of the problems found in gathering in this field spring from the bundling process. As magazines and paperbacks are often much thinner than books many companies find it possible to dispense with this operation and the sections arrive at gathering in best condition for fast output. Mis-feed detectors are incorporated with an optical warning method for locating the fault. When the gatherer stops the control system allows the books in the binder to run out but should the binder halt then the preceding gatherer will also come to a stop.

Successful adhesive binding requires the correct balancing of four variable factors and considerable experimentation is proceeding to try to improve the strength of the bonds obtained. In the production situation the binding-machine operator has no control over the type of paper on which the job is printed and permutations on the various methods of spine preparation and adhesives that can be used must start from an understanding of the construction of paper stock. The other invariable factor is the nature of the machine itself. Although certain units on the chasis may be changed or modified each machine type has its own characteristics which have to be taken into account at the time of purchase.

Spine preparation

Machines are installed with the type of spine preparation mechanism the manufacturers think appropriate and these include combinations of rotary knives, milling cutters and grinders, thin bladed and circular saws with sanding and scoring mechanisms as subsidiary treatments. All these combinations tend to give different strengths to the subsequent bond and a thorough understanding of the principles involved is needed to produce the best bond in a given situation.

The earliest method of spine removal made use of a simple rotating slitter wheel that could be sharpened *in situ*. This worked against a pressure wheel which helped to prevent distortion of the spine as it was subjected to very large pressure forces. This method was found to be adequate while the stock being bound was of the porous unfilled variety, while speeds were slow and aqueous adhesives were used. Sometimes a supplementary sander might be used to fibrillate or tease out the fibres on the spine to anchor them in a film of adhesive.

A second method is the use of a raised sharp steel point or stylus to score across the spine and produce microgrooves. The point is often set in the centre of a cutter which has a number of carbide tipped teeth. Working through a radius of about 30mm it produces closely set multi-directional

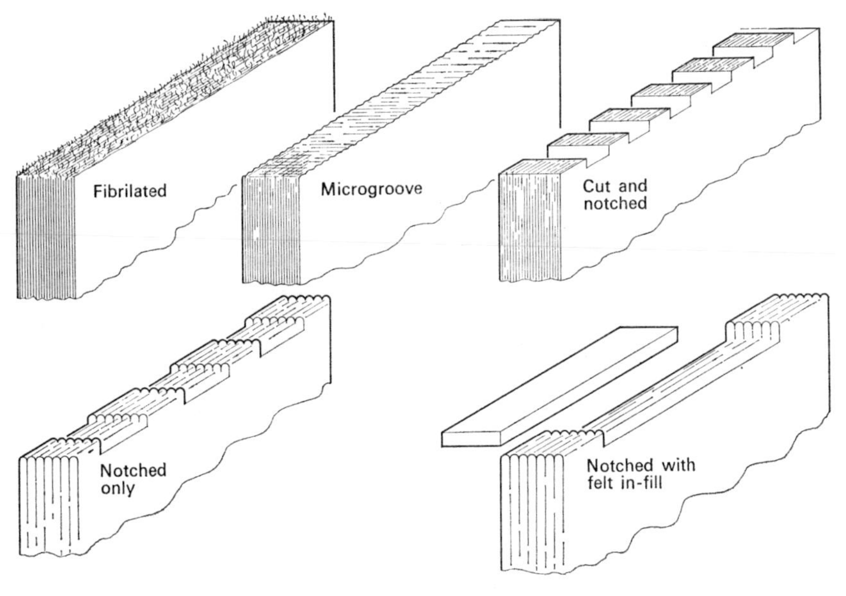

15.1 *Variations in spine preparation.*

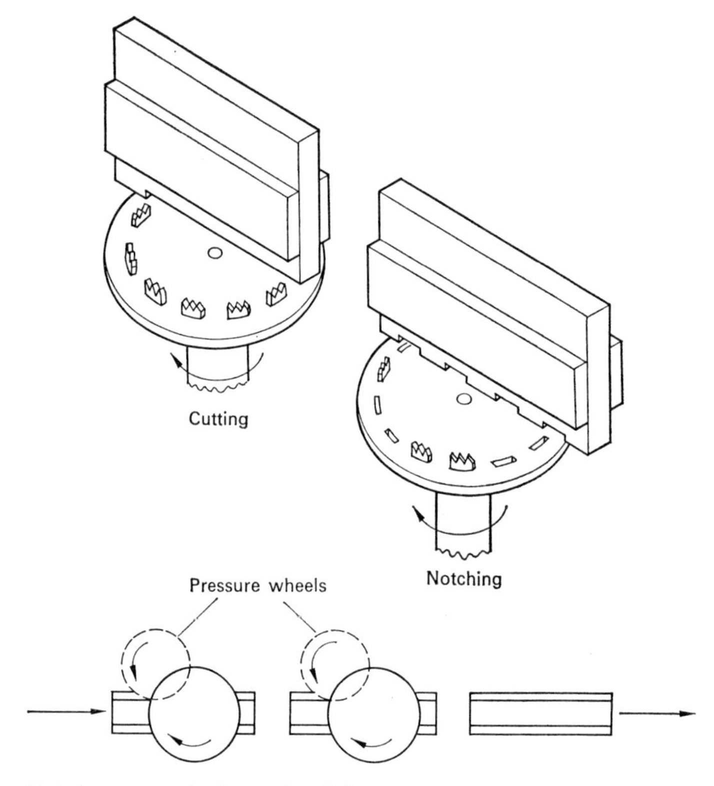

15.2 *Carbide cutters may both cut and notch the spine.*

grooves that both increase the surface area exposed to the adhesive and introduces fibrillation in some stocks (fig. 15.1).

The most recent of the methods introduced utilises carbide tipped cutters that produce patterned roughing or notching across the spine. These are so set that the cutting action is uni-directional and produce cuts up to 10mm apart (fig. 15.2). The theory of this is that because the 'tabs' are long they are embedded in the glue film and shift the bending stress on to the leaf when in use. Books with a flat spine tend to hinge from the glue line and this is where a large part of the deformation occurs. By pattern notching the tabs are in a glue film and flexing occurs along a line connecting the base of the notches. The American engineers W O Faxon and R E Fogg illustrate a method of achieving a pattern of this type in their paper 'Mechanics of adhesive binding'.

It is clearly advantageous to be able to vary the rotary speed of the milling head in relation to the binder clamp velocity and one machine offers speeds of

450, 600, 750, 1 200 and 1 500 revolutions per minute whilst a faster machine requires speeds between 3 600 and 7 200 rpm. A typical speed for micro-grooving would be about 2 800 rpm.

Experimental methods of spine preparation include one technique of notching into a book that retains its spine folds. The deep notch obtained is then filled with a soft felt pad and the book bound in the usual way. By leaving the head and tail folds intact it is hoped to prevent the breakdown of the spine bond which often starts at these points.

Work done by C V Hawkes of PIRA (*Printing Journal*, Spring 1968, Vol. 1, No. 1) concludes that some increases in page pull strength may be obtained on all classes of stock by the use of controlled close notching, but most gain may be on coated stock. Direction of grain does not significantly alter the value obtained except in the case of newsprint and super-calendered stocks which appeared to be strongest with paper grain at right angles to the spine. Another factor is the provision of good machine brushing and dust extraction facilities which is vital to the successful binding of coated stock. The milling and other preparation previously discussed tends to release the chalky materials used as a fine powder and this inhibits the 'wetting' of the surface if not removed.

Adhesives

When animal glue is used for adhesive binding it is a specially formulated elastic and high tacking version. It is very popular for the production of thick books such as telephone directories, timetables and similar work printed on stock of a porous nature which allows deep penetration of the glue into the fibre. Coupled with the thin flexible nature of the paper and the mull lining between the book and its cover a relatively strong attachment is obtained against a low adhesive cost. As there must be a drying period between binding and trimming some large installations have a drying conveyor which delays the delivery of the book while this occurs.

The viscosity of all heated adhesives varies with temperature and to obtain maximum penetration into the valleys of the roughened spine it is desirable that the adhesive should be applied at the highest temperature possible consistant with the maintenance of quality. A method of lowering the viscosity of glue is to add water but this also reduces tack and slows the drying and in this instance is highly undesirable.

Polyvinyl acetate is also specially formulated for the adhesive binding machine to provide correct initial tack and setting speeds. Machines using this adhesive usually run at speeds up to 3 500 copies per hour with very much lower outputs for thicker work. Delivery is usually to a flat platform from where manual removal to drying racks is necessary. A minimum drying period of two hours is needed and many users find it desirable to leave work for 24 hours before attempting to trim the edges. This involves the storage of many thousands of wrappered books in special containers and a large area of the bindery must be given over to this purpose. This has been the major factor in the changeover from PVA to hot melt, the companies being able to economise on space and implement true in-line sequence production. Nevertheless polyvinyl acetate has some advantages over hot melts and these have prompted the continued use of this adhesive, economies being obtained by speeding the production with post-binding drying mechanisms. The advantages are cheaper adhesives, greater flexibility in the finished product and better bonds on certain classes of stock.

The introduction of hot-melt adhesives has revolutionised the productive capacity of adhesive binding machines which now run at speeds up to 12000 copies per hour, and can be linked directly to the trimming operation, shortening production times and saving on space. Formulations of hot melts specify the working temperature and accuracy between ± 5°C is usually required. Heating above this usually involves the risk of overcooking the adhesive, causing severe degradation and spoiling. Application of the adhesive at lower temperatures increases the viscosity and reduces its ability to flow into the prepared spine. The relatively simple applicator pots fitted to small machines tend to develop hot and cold areas of adhesive and to overcome this the rollers of the more sophisticated installations are heated. This ensures that at the point of application the adhesive is at the highest temperature possible and helps to prevent undesirable variations.

Examination of plan illustrations of adhesive binding machines shows that the point of adhesive application is from one to ten stations or clamps from the cover attachment which in turn is a similar distance from the delivery point. It is necessary for the adhesive to have reached its maximum tackiness before the attachment of the cover and to be set off between cover attachment and delivery area. The 'open time', *ie* time between application and setting is therefore critical for each machine and speed. To illustrate this point consider the 28 clamp machine in figure 15.3. A running speed of 150 books per minute equals 2½ clamps per second; with 6 clamps to the creaser and 14 to delivery the open time should be between 2 and 5 seconds. At a slower speed of 60 books per minute, *ie* one clamp per second, an adhesive open time of 6 to 13 seconds is required.

New principles being evolved for applicator pots for hot-melt adhesives incorporate two units; the pre-melter which is connected by heated lines to a much smaller applicator. The first unit pre-melts the adhesive at about the rate it is being used and it is then pumped to the point of application. The adhesive is kept moving to prevent the hot spots previously mentioned and the transfer surfaces are heated. Any adhesive not lifted off the roller by the book spine is re-circulated to the pre-melter and brought back to correct temperature. One of the main advantages claimed for this particular system is that because of its size it is possible to mount the applicator closer to the covering station than was previously possible. The pumping of the adhesive through the system makes for flexibility with the pre-melter sited to suit the installation.

The glue film applied to the book spine is metred to suit the requirements of the job. By careful measurement the film can be reduced to an optimum point where maximum strength is being obtained for thickness of adhesive applied. A thicker application would simply use more adhesive, and therefore cost more, without necessarily improving the bond to any significant degree. Methods of controlling the glue film include doctor blades and metering rollers but to be entirely successful the system should also include a method of adjusting the vertical position of the applicator rollers in relation to the spine.

Machines

The path of the book through the adhesive binding/covering machine may be a straight line, circular or elliptical. Generally speaking the straight line machines have the lowest productivity and the elliptical versions the highest; cost also tends to run along these lines. The small straight-line machines are usually hand fed and take-off and are worked as individual machines without any linkage to other units.

Both the circular and the elliptical types may be hand fed but are increasingly linked into more complex systems. As the principle of adhesive binding is being applied to a wider range of work than hitherto so machines are being offered with a wider range of possible product. Some of the larger machines are able to accommodate four distinctly different types of work: 1, straightforward wrappered (or covered) work with or without on-machine creasing of the covers; 2, wrappered with spine mull strip; 3, lined with calico or paper strip and side glued; 4, lined with mull and paper strips both sides glued. A machine with this potential could be used on a wide range of sewn, stitched or adhesive bound paper-covered work and book blocks for subsequent case binding.

Transfer of the book from the gathering machine to the open clamp of the binder is achieved along a raceway that is curved to turn the book from a horizontal to a vertical position. When necessary a side wirestitching unit is sited midway along the raceway to stitch the occasional job that cannot be adhesive bound. This usually comprises two double-headed units operating at half the binder speed. Books are delivered from the raceway up into the open clamp where they are jogged before the clamp closes. Hand feeding is normally possible and provision is made for up to three operatives so that binding speeds remain high. Thread-sewn work is introduced into the machine in this way.

Good spring pressure on clamps is important, as the stock must be held tightly whilst the book is running through the knife or grinder. Any tendency to tip the book will cause misalignment in relation to the subsequent glue roll and consequent bad adhesive application. As clamp pressures tend to be relatively low it is important that minimum force is used by the spine preparation method. The spine preparation unit is sited immediately following the clamp closure and is linked to a suitable suction head to remove the swarf. Where possible the spine is also brushed to remove surplus loading or coating material, and the waste is passed to a collection box or bag.

The gluing mechanism immediately follows the spine preparation and this allows the maximum time lapse for the adhesive to tack off before cover attachment. In the case of the 'two shot' system the PVA application is followed by a line of infra red heaters or a dielectric heating field and a hot-melt applicator is sited closer to the cover-attachment point. Those machines that have the facility to apply mull or other lining materials have the reel-feeding, cutting and application unit stationed next. As the material is cut from a reel that is as wide as the book is long, the strip, when applied has cross-grain direction. This can cause difficulties when aqueous adhesives are used and even relatively stable crêped kraft paper may develop unwanted kinks and blisters.

Both pile and continuous types of cover feeders are used on the machines and in some instances the cover can be passed between creasing rollers before being attached to the book. The transporter chain is carefully timed to deliver the cover in register flush with the head of the book. It is usually desirable for the cover to be about 3–4mm long at the tail to ensure that surplus adhesive does not flow from the spine on to the working surfaces of the machine. From the moment at which the book and cover are in contact the two are kept under constant pressure along the spine to ensure good bonding between the two units. To obtain a neat sharp finish to the spine corners the cover is broken over by the pressing station or stations. This also ensures that the book block is consolidated and that any adhesion between cover and side gluing is confirmed. Facilities are also available to ensure lateral register of the cover on the book block.

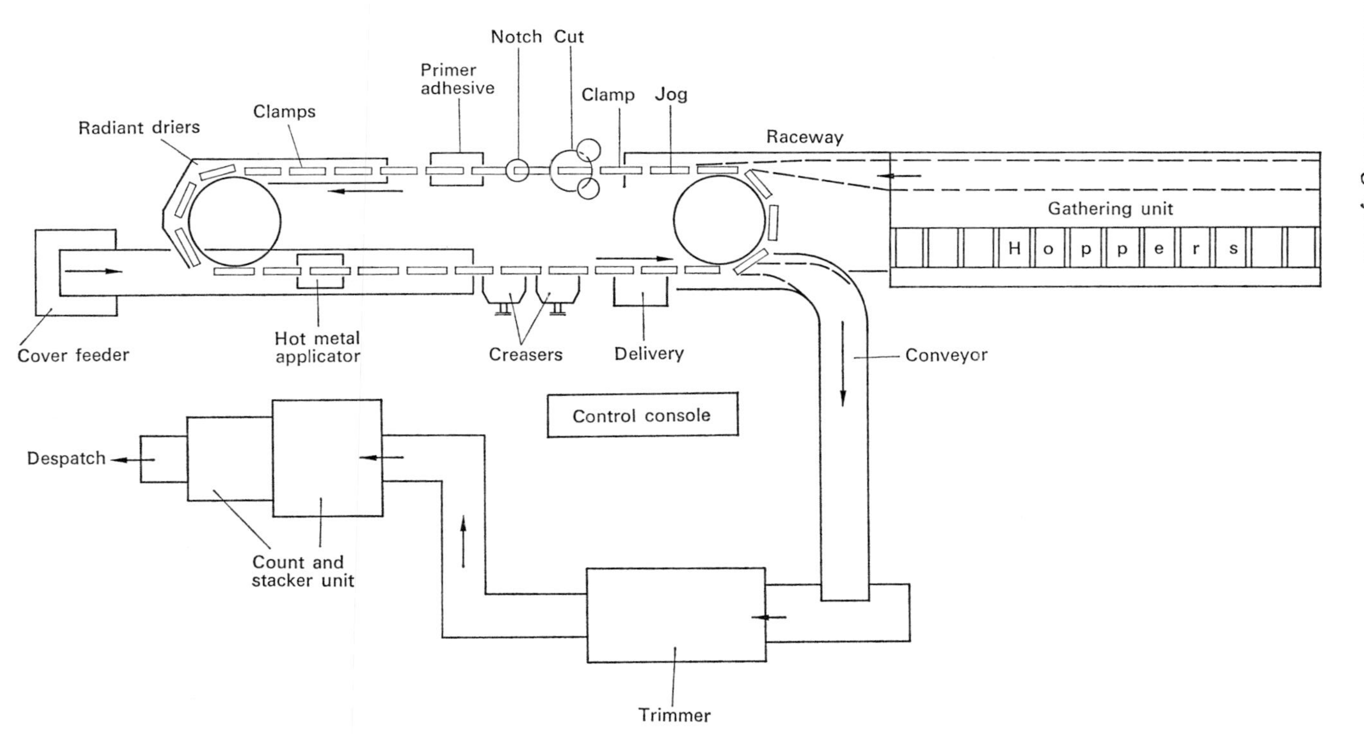

15.3 *Plan of two-shot binder linked to gathering, stacking and trimming units.*

At the point of delivery, work that has been bound with hot melt may be passed directly to the trimming machine, but when the slower-drying PVA has been used very careful handling will be required. A sound method of drying PVA work is to stack the books back down on a shelved trolley for about twelve hours, but of course this is slow. Larger installations utilise heated pocket driers or a clamp system that passes the work through a dielectric drying system before delivery.

As in the case of saddle-stitched work the dry adhesive bound work may be trimmed singly or in piles and to some extent the size and speed of the binding installation and its product will control which method is adopted. The slower installations can be serviced by the single-book trimmer, versions of which will operate up to around 80 books per minute and to 40mm thickness. Binding speeds beyond this are quite common, particularly on the thinner magazines and paperbacks and it is usual to assemble these in suitable piles before trimming on a three- or five-knife machine. It may be necessary to introduce a unit between the binder delivery and the trimmer to perform the function of piling, but some of the more sophisticated trimmers have the ability to select piles of the correct height from an assembly hopper on the machine.

One difficulty arises in trimming work that has been printed, gathered and bound two up is that of cover-break on the spine. Although many three-knife trimmers have the potential of splitting two-up work and returning the tail half to the operator for retrimming, the thickness of the tail knife will create a bevel to the outside half of the cut and nearly always distort or split the covers on the spine. Although one very sophisticated machine has a patent 'back grooving' attachment, nearly all manufacturers of paperbacks prefer to split the two-up work with a band saw or circular saw mechanism before treating each half individually on three-knife trimmers.

Another feature of the production of paperbacks is the use of the so-called 'come and go' imposition, and this complicates the division of the work still further. This work is imposed either head to head or tail to tail, so that viewed from the face, the top book shows page one of the cover and the bottom book shows page four of the cover. This implies that, after splitting, one half has to be turned over before trimming and this requires another piece of mechanism between splitter and trimmer.

Quality control in adhesive binding

Apart from the obvious visual checks for cover register, squareness, size and general professional finish, two methods of testing strengths of adhesive bindings are used, *ie* page flexing and page pulling. The former method has not been pursued fully because of the rather lengthy testing times involved. The book is fastened and the leaf to be tested is placed under tension in a clamp that has a reciprocating action. The leaf is then flexed back and forth until is becomes detached, the number of movements necessary giving a measure of the strength of bond. Most leaves require several hundreds of flexing movements before detachment occurs.

A more usual method is a tensile test aimed at measuring the pull necessary to detach the leaf from the glue film and many companies use the PIRA page puller for this purpose. The machine consists of a clamp bed to which is fixed the cover and five leaves at each end of the book to be tested. The sixth leaf is then locked into a moveable clamp that is made to rise under spring pressure. The small geared electric motor providing the power continues to exert pull through the springs until the leaf is detached (fig. 15.4). The power used, in kilograms, can then be read off on the gauge, and by taking readings at regular intervals throughout the book an average is obtained. Although

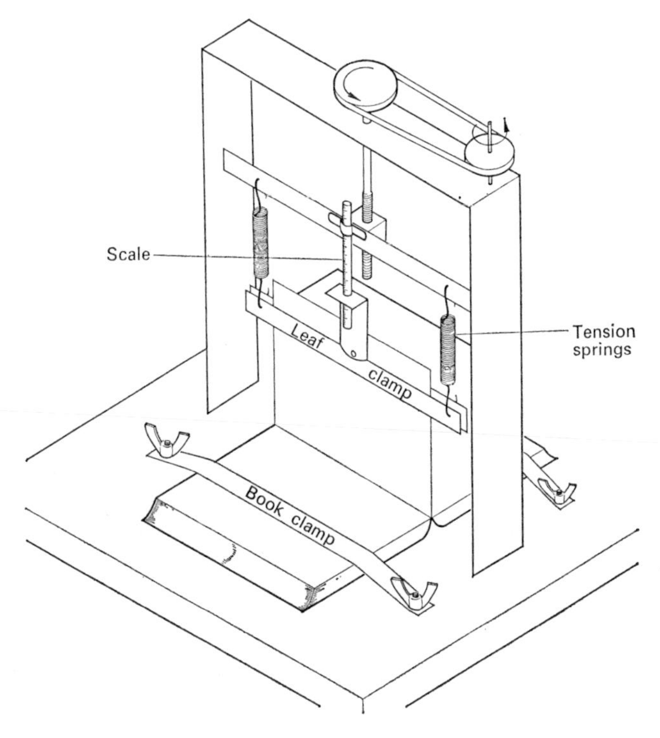

15.4 *A single leaf is pulled by the* PIRA *page puller to test the adhesive bond strength.*

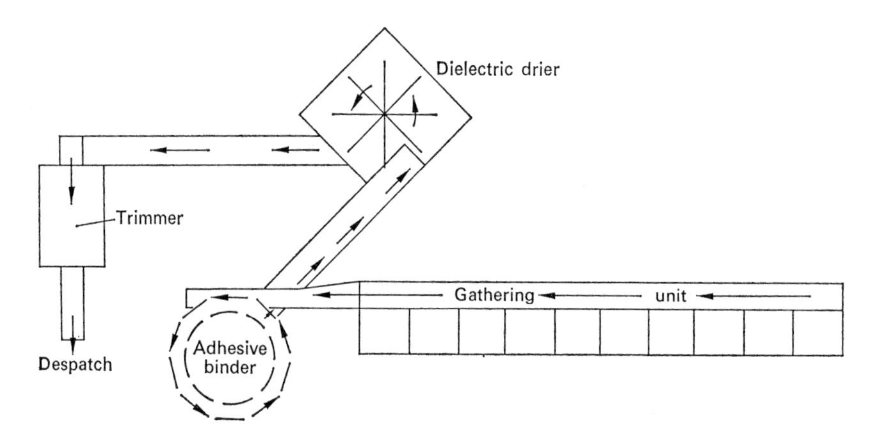

15.5 *Adhesive binder combination incorporating gatherer, circular binder, carousel-type dielectric drier and trimmer.*

very few figures have been published some manufacturers are content with around 6 kg per 100 mm of book length for coated stocks and 9kg per 100 mm for uncoated stocks.

Developments

All the present signs are that in the foreseeable future more and more traditional bookbinding areas will succumb to the adhesive binding process. Already most children's toybooks, many novels, book-club and presentation work is bound this way; as improved methods are developed and adopted the economics of the process will tend to make it more attractive to the publishers.

One interesting rotary letterpress printing machine that can produce books instead of sections bids fair to introduce the automatic book production unit of the future. On this machine paper reels are printed, on both sides, with a complete book and the reel is perforated, slit, collated and delivered in alignment. The rubber printing plates are laid up long grain, in the direction of web travel, producing curl-free long-grain pages. Books 150 × 110 mm containing 320 pages issue from the press at 100 books per minute ready for immediate insertion into the adhesive binder.

16. Organisation and workshop layout

In common with other manufactured articles, paper products have a natural production sequence. If a job is to move smoothly through production then the machines and other processing areas must be located in positions which will allow each operation to be completed in correct order. Some machines are themselves capable of several separate functions or may be connected in-line and the entry and exit of these units must be correctly placed in the general flow. The shape of the production line is nearly always governed by the geography of the building; it is seldom possible to plan the building around the line. Critical factors in the layout of a workshop will include such points as location of entry and exit of raw materials, location of doors, fire exits and windows; physical shape of the building, *eg* single or multi-floor; whether special light is required etc.

The nature of the machines to be used must also be considered; flooring may have to be reinforced if heavy machines are installed above ground floor level. If heavy raw materials are to be used, *eg* strawboard, it may be possible to minimise the distances stock has to be moved. Decisions on how the stock and work in progress is to be moved around in the workshop are important too; fork-lift trucks require wider gangways than simple hand-operated hydraulic trucks.

Every factor of production should be studied and understood if a successful layout is to be constructed and examples appear for a small miscellaneous bindery (fig. 16.1), a print finishing department (fig. 16.2) and for a mass production publisher's bindery (fig. 16.3). In each case a relatively simple production basis has been taken although in practice departments are complex in character and a particular job may have to follow a route that is unsatisfactory. This may be especially true in a small organisation having only one of a vital machine, *eg* a guillotine, that is really needed at several points in the production line. In this case the choice of location is a compromise and may suit only a percentage of the jobs passing through.

Production engineers have techniques to explore the route that a job takes on its way through a workshop; one such method is described as a string

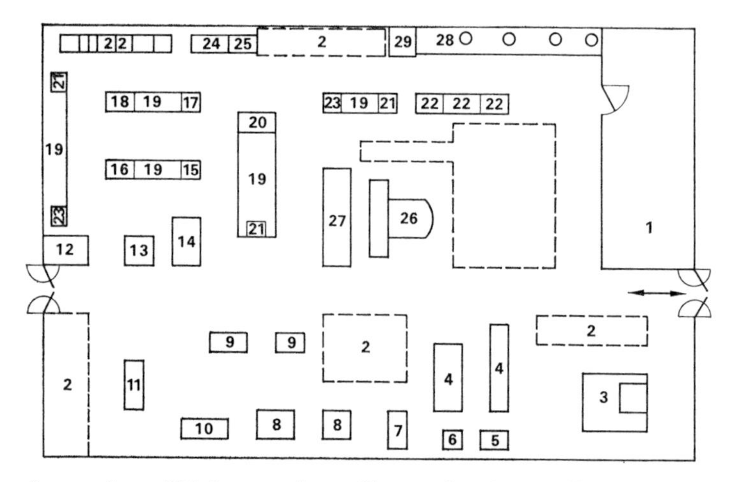

16.1 *Layout of a small bindery to produce a wide range of stationery and letterpress books.*

1 *Materials store*
2 *Work in progress*
3 *Hand-fed ruling machine*
4 *Benches for women's work*
5 *Treadle punch*
6 *Power drill*
7 *Treadle perforator*
8 *Sewing machine*
9 *Wirestitcher*
10 *Creasing machine*
11 *Stripping machine*
12 *Glue kettle*
13 *Hand board cutter*
14 *Hand-operated guillotine*
15 *Leather paring machine*
16 *Nipping press*
17 *Roller backing machine*
18 *Rounding machine*
19 *Benches for male work*
20 *Sheet gluing machine*
21 *Bench mounted press*
22 *Standing press*
23 *Glue pots*
24 *Numbering machine*
25 *Indexing machine*
26 *Power guillotine*
27 *Guillotine bench*
28 *Finishing bench with stoves*
29 *Blocking machine*

diagram. This requires that a scale plan of the proposed layout be prepared and a pin located at each process stage. By drawing a string around the pins from stage to stage in the path a particular job takes, the route can be clearly seen; further, by removing the string a direct measurement of the length of the path can be made and this may then be compared to a similar measurement on an alternative layout. By winding the paths of several jobs on to one diagram, in different colour cords, a measure of the likely congestion areas or bottlenecks may be indicated. Of course, it must be appreciated that this device is concerned with space and does not take time into account.

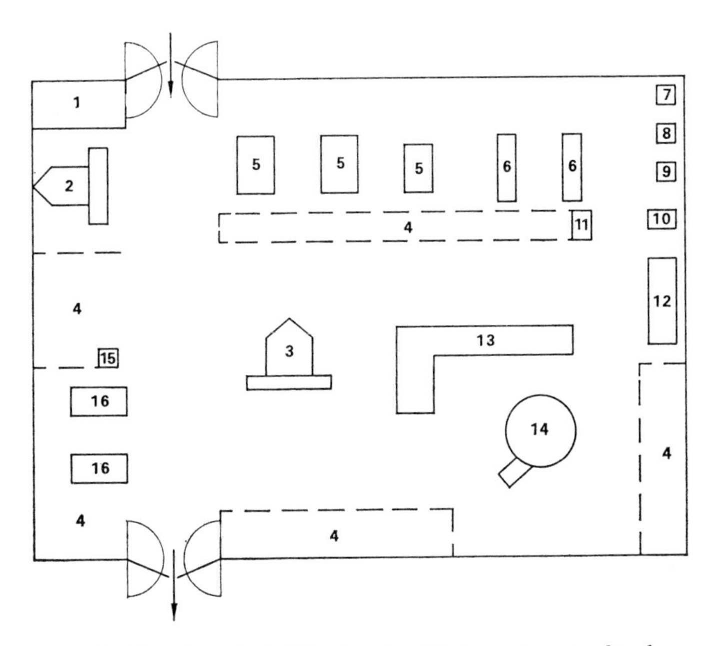

16.2 *Layout of medium-size print finishing department for processing cut and pack miscellaneous work; labels; brochures; medium-length runs of inset work; square-backed adhesive bound or wirestitched wrappered work.*

1 *Store*
2 *Programmatic guillotine*
3 *Standard guillotine*
4 *Work in progress*
5 *Folding machines*
6 *Make-up benches*
7 *Heavy wirestitcher*
8 *Power paper punch*
9 *Power drill*
10 *Perforator*
11 *Standard wirestitcher*
12 *Miscellaneous machines*
13 *Insetter-stitcher*
14 *Small carousel-type adhesive binder*
15 *Paper counting machine*
16 *Packing bench*

This technique has important uses in the miscellaneous workshop, *eg* bindery or manufacturing warehouse where a wide variety of short-run jobs are processed and involve the use of innumerable small machines. By relating the string diagram to a careful study of the nature of the work processed a balanced main path may be determined that will perhaps satisfy the bulk of the work. Investigations of this sort may reveal that considerable savings in work movement may be made if the processes are arranged in a particular way and sometimes show that the placing of a small piece of equipment or operative in a vital position may considerably simplify the path. Moveable benches, machines and racking all help to improve the flexibility of the workshop and assist in redesigning the flow when this is desirable. Machines can only be repositioned easily if the electrical services are also flexible and modern factories may have installations that allow for this, *eg* the overhead 'bus' bar system.

Due attention must be given to the provision of areas set aside for the storage of work in progress. This is particularly relevant in print finishing processes, as it is often necessary to partly process a job and then to stack it for drying or to await the availability of other machinery, *eg* building up of a sufficient quantity of machine-sewn books before attempting to forward them. Sometimes conveyors can be used between processes and by suitably adjusting the speeds of the conveying unit storage areas are created that stockpile the work in progress and take up variations in output between machines.

In very large adhesive-binding installations a buffer system is made available to provide a reservoir of books between the gathering and binding units. The reasoning here is that although the binder will work continuously, the action of the gatherer tends to be intermittent. When the two units are linked, the binder also has to work in an intermittent fashion and this may spoil books on the machine and place the mechanism under strain by constantly stopping and starting. The system requires that the gatherer first fills the reservoir with books; the binding machine is then started, receiving its supply direct from the gatherer. If this fails to provide a continuous level of work,

---

Key to layout of publishers' bindery (see fig. 16.3).

1 *Quad folding machine*
2 *Combination folding machine*
3 *Bundling machine*
4 *Work in progress*
5 *Guillotine*
6 *Endpapering machine*
7 *Plating and make-up bench*
8 *Gathering machine*
9 *Sewing machine*
10 *Conveyor belt*
11 *Assembly of sewn books*
12 *Nipping machine*
13 *Book back gluer/drying machine*
14 *Three-knife trimmer*
15 *Edge gilding machine*
16 *Edge colouring machine*
17 *Sewn or adhesive calico lining machine*
18 *Automatic rounding and backing machine*
19 *Lining machine*
20 *Casing-in machine*
21 *Book forming and pressing machine*
22 *Jacketing machine*
23 *Rotary board cutter*
24 *Cloth cutting machine*
25 *Sheet-fed casemaking machine*
26 *Automatic platen blocking machine*
27 *Four-pillar blocking machine*
28 *Inspection and jacketing bench*
29 *Packing benches*
30 *Area for the production of samples, dummy copies and for the repair of spoilt work*
31 *Board, cloth and jacket store*

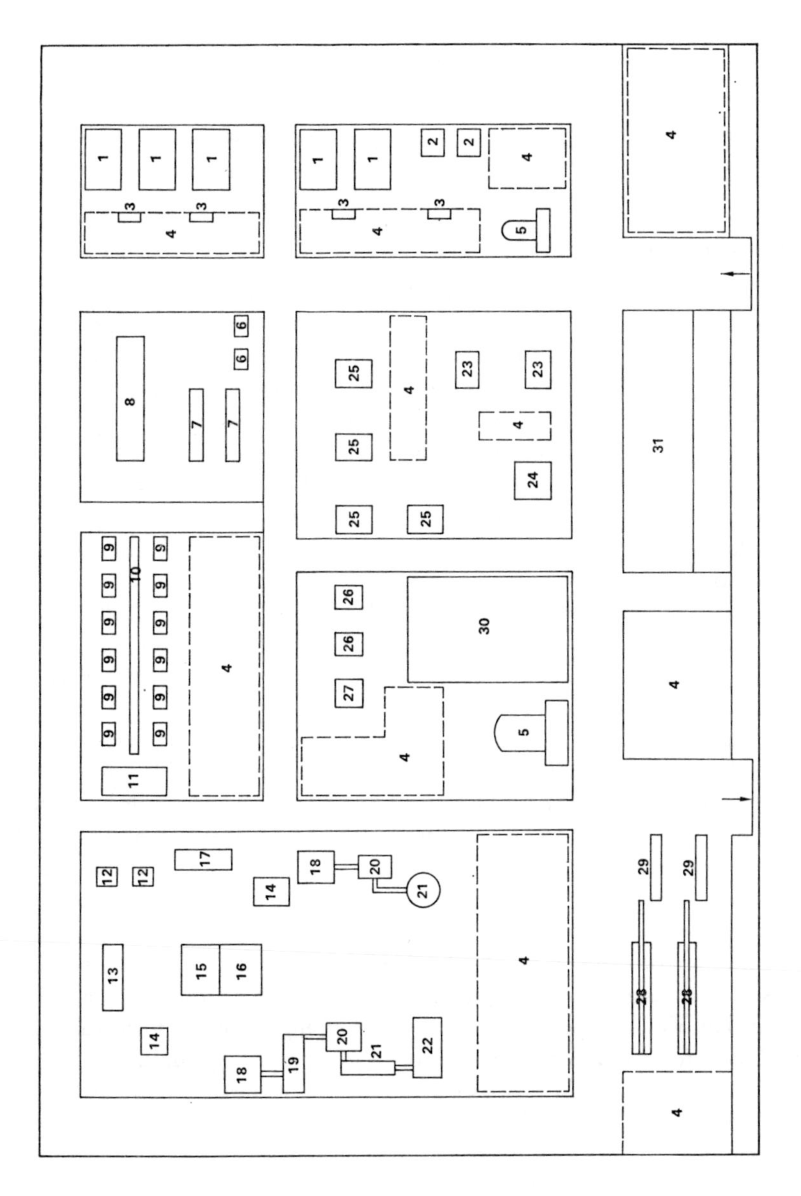

16.3 *Layout of a publishers' bindery for the production of standard-size sewn and unsewn books (see facing page for key).*

the reservoir automatically switches in and tops up the supply to the required level. The speed of the gathering unit is then slightly increased to refill the reservoir and is slowed to binder speed when the reservoir rails are full. In this way minor stops (up to 10 seconds) in the gatherer are not felt at the binding stage.

PRODUCTION CONTROL

The application of this management technique will depend upon the size of the company and the nature of its product. The object is to make the best use of the means of production so that job delivery dates may be met, thereby improving the service to customers and in so doing increase sales and profit. By accurately planning ahead bottlenecks and unnecessary spare operational time can be foreseen and perhaps avoided; the installation of such a system will involve management in an expense that will have to be recovered and the savings must be sufficient to offset this. Many managers of small departments complete their planning on a hand-written planning schedule, but these are often not available to other members of the production staff and for larger establishments a more positive and visible system may be desirable.

Preparation of such a system requires that full details of all machinery is known, *ie* capacity, speed, limitations, etc. To arrive at a correct assessment of true production times for given types of work it is usually necessary to study each phase of production over an extended period. The data derived from these studies can then be used to compile detailed tables from which future production times may be calculated. Alternatively production times may be extracted from costing system documents. Initial analysis may be spread over a considerable period of time and even after installation of the control system time studies will continue to be made to ensure that data is checked and kept up to date.

Although production may be divided into natural departments, *eg* folding, make-up, sewing, forwarding, casemaking and despatch, each will have to be divided into suitable production units or groupings; *eg* in the forwarding area the linked forwarding machines can be considered a single unit. At this stage it is possible to calculate the available capacity of these 'loading areas', due consideration being given to shift work or regular overtime that may be worked.

Presentation of the recorded data is usually made on a board that shows clearly the weeks ahead, the loading centre, maximum loading possible and the current situation. By utilising a system of colour coding the board can indicate various classes of work and be labelled with job numbers or titles;

DAY WORK CHART

Machine		Week No. / Day No. 21/1	21/2	21/3	21/4	21/5	21/6	Week No. / Day No. 22/1
Folding Mc. A	1	106						
B	2	106						
C	3	38	38					
Insetter·stitcher·trimmer	4							
Guillotine A	5		106		38			
Prog. B	6	216						216
Miscellaneous Mcs.	7		106					
Adhesive binder	8			38				
Female benchwork A	9	106		38				
B	10							
C	11							
Male benchwork	12							
Despatch A			216, 106		38			
B								

16.4 *Production control chart for daywork (6 days) showing three jobs programmed.*

Job 106	*Fold 16pp (5 hrs)*	*Mc. 1*	*Job 38*	*Fold (16 hrs)*	*Mc. 3*	*Job 216*	*Regular weekly job*	
	Fold covers (2 hrs)	*Mc. 2*		*Gather (10 hrs)*	*Mc. 9*		*Trim on programmatic*	
	Inset by hand (2 hrs)	*Bench 9*		*Adhesive bind (6 hrs)*	*Mc. 8*		*guillotine (5 hrs)*	*Mc. 6*
	Wirestitch (3 hrs)	*Mc. 7*		*Trim (6)*	*Mc. 5*		*Despatch (2 hrs)*	*A.*
	Trim (guillotine 3 hrs)	*Mc. 5*		*Despatch (4 hrs)*	*A.*			
	Despatch (2 hrs)	*A.*						

if various lengths of indicator are used an idea of the total time element is also given. How far forward planning is taken depends upon the class of work going through the works and may be for two, three or more weeks ahead. A wide variety of commercial boards are available, but small companies may decide to use a simple chalkboard or printed chart and although these are cheaper and easier to maintain they are also less comprehensive (fig. 16.4).

Once the production-control system is in operation it is necessary to ensure that it functions correctly and that jobs are progressing satisfactorily. This is the purpose of the progress clerk, who will correlate information about completed work in relation to the planning board; any hold-ups will be revealed and action taken. Jobs having a firm delivery date may be programmed from the despatch department back to the commencement of the work. In this way the minimum time the job will take to pass through the workshop will be shown and provide the latest date for receipt of sheets, any outwork or materials used.

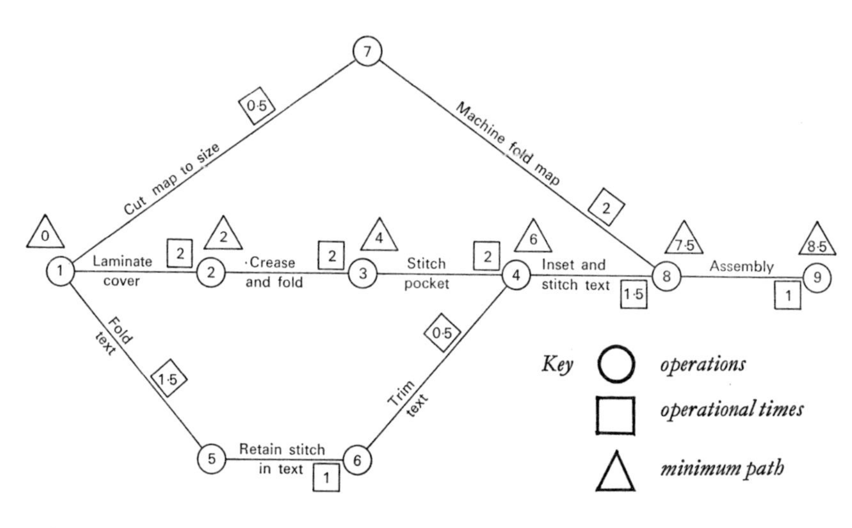

16.5 *Network analysis (critical path) for a finishing job.*

A technique that may be of value in assessing the time factor in more complex jobs is called 'network analysis'. This is a chart or model showing the path through which the elements of a job pass. By measuring the lengths of the path in hours, the minimum length of time needed to complete the job can be calculated. Figure 16.5 illustrates a simple chart for a wirestitched booklet with the following specification: 32 pages saddle stitched into laminated and overhang cover. The cover has two wirestitched pockets closed by wirestitches, one containing a 24 page prospectus folded map. Time for operations are laminate 2 hrs; crease and fold cover 3 hrs; stitch pockets 3 hrs; fold 32 page $1\frac{1}{2}$ hrs; retain stitch 1 hr; inset and stitch into cover 2 wires $1\frac{1}{2}$ hrs; cut map to size $\frac{1}{2}$ hr; fold map 24 page prospectus 2 hrs; assembly 1 hr.

Trends and developments in finishing operations

The rapid technological advances being made in other sections of the printing industry have not yet appeared in the finishing field. Although the newer machinery being installed may embody many engineering and electronic refinements the basic methods of construction and production of finishing processes remain surprisingly unchanged. The principal improvements to be seen are those concerned with systemising production to achieve better workflow at higher speeds and the use of multi-process link-ups. However, machinery of this type often has a limited range of abilities and sizes with a consequent loss of flexibility of product; firms that utilise this equipment often have to specialise rather heavily.

Examples of multi-process machines in various sections are as follows:

Print finishing

Insetter–stitcher–trimmer combinations with a predetermined number of feeder heads. Inserting and mailing equipment may be attached in-line.

Stationery work

Automatic reel-fed machines to rule, sheet, count, add cover and board back, punch and spiral-bind reporters' notebooks and other simple styles.

Reel-fed ruling machines that rule both sides, print cross-lines from rubber stereos, fold, add pre-printed cover, wirestitch and trim student exercise books at high speeds.

Paperbacks and squareback magazines

Automatic gathering, binding, trimming, inserting and shrink-wrapping machines in-line. Soon it may be possible to dispense with the gathering process; a new rotary printing machine has been developed that issues books instead of sections and as this gains favour certain classes of adhesive-bound work may become automatic from the reel (fig. 16.6).

With the exception of these and a few other examples the application of true automation is seldom possible; however, with the rationalisation of the industry now in progress the product, too, may be rationalised and longer more specialised runs will allow better use of completely automatic lines.

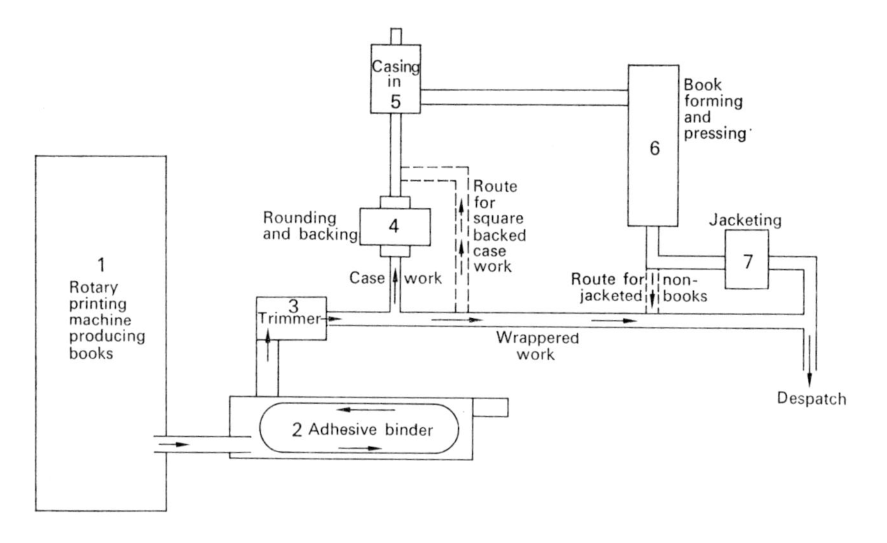

16.6 *A theoretical layout for the production of both wrappered and case-bound books by the unsewn process. Machines in the layout include:*

1 *Rotary belt printing press*
2 *Large multi-purpose adhesive binder*
3 *Trimmer*
4 *Rounding and backing machine*
5 *Casing-in machine*
6 *Book forming and pressing machine*
7 *Jacketing machine*

QUALITY CONTROL

Specific departments concerned with quality control are confined to larger companies, but the function must appear in all manufacturing organisations. Theoretically it is ideal if all component units and materials used in production are checked upon arrival at the factory and each subsequent usage checked against some form of control. Paper and paperboard may be inspected this way if sufficient quantities are purchased to make it a viable operation. Tests may be run to confirm the manufacturer's specification and to gauge the qualities it may have in relation to printing, creasing and other manufacturing processes. When large quantities are involved consistancy of quality from batch to batch is a particularly important aspect. Testing of paper and paperboard is very well established and documented in the literature available from the national research organisations.

Few companies find it possible to run departments to control the quality of incoming binding and finishing materials although many tests are possible. Sometimes materials are pre-tested before a run commences by the manufacture of dummy or sample copies in very small quantities by semi-hand methods. However, quite different effects and reactions may be the case when the job is running through high speed machinery. Tests on binding materials may include:

Endpapers

Dimensional stability in relation to aqueous adhesives, colour fastness, caliper, grain direction.

Cover paper

The above plus surface scuff resistance; if printed the quality of the printing in relation to scuffing and rubproofness; if laminated the fold stiffness and cohesive quality of adhesive to withstand joint forming under heat and pressure.

Textiles

Visual inspection of surface for regularity of finish and colour, abrasion and flexing qualities, resistance to vermin attack.

Boards

Dimension, caliper–weight relationship, moisture content, flatness (visual).

Adhesives

pH, initial tack, flexibility, odour, physical condition, viscosity.

Foils

Colour, suitability for job surface.

Ensuring that each phase of production is correct is usually the responsibility of departmental management who, themselves, may initiate some quality control process. Where the article produced is an assembly, *eg* a book, calendar or showcard, the various units are manufactured to the highest standard of accuracy so that integration can occur with minimum difficulty. For instance, finished size is an important factor and may, indeed, be vital if several units are to be assembled into, say, a loose-leaf binder; both the size of sheets and the punching must be accurate if the sheets are to align neatly in the cover. Accurate size has particular relevance if the paper or paperboard article is subsequently to be used on sophisticated machinery, *eg* punched cards on computers, cartons on package-erecting machinery and labels on canning and bottling lines. Engineers in these fields rightly require the highest standards of accuracy obtainable and tolerance of ± ·025mm are frequently demanded.

Many finished articles have both a visual and a functional aspect. A simple example is the ability of showcards to stand; the position of the stand/hanger on the back is critical if the angle is to meet the requirements of the customer. Other examples are the functioning of an account book, the slide action of a carton and the fit of a box lid. The appearance of completed work can easily be spoiled by a simple act in the finishing department and may lead to the lowering of the overall quality of the product. Slightly excessive pressure on a guillotine clamp may show as a hard line on the top copies of a pile and can usually be avoided by reducing clamp pressure or masking the job with some waste sheets. Showcards with border designs require very careful production if the visual effect is to be satisfactory. Book cases made from pre-printed paper felts require the printing to be accurately positioned in relation to the hollow of the book so that lettering can be seen to be nicely placed on the spine.

These and many others are the areas of inspection that may be checked by the quality control department; although departmental management may have passed the job for running it does not follow that consistent standards will be maintained.

An element of finishing production that is not easy to test during production is the success of bond obtained with adhesives. It may be necessary here to test objectively and regularly so that trends and patterns become clear. As finishing adhesives may have cohesive strengths greater than the tensile strength of the papers being processed, bond failures are rare. But nevertheless in packaging, publishers' binding, loose leaf covering and in multi-unit display assemblies failures do occur due to poor bonding and these are not usually revealed until the job is completed or is in the field. Regular tests may expose the reasons for this and appropriate steps taken.

Reasons for failure of adhesive bonds include 1, using adhesives with insufficient cohesive strength for the task, due to dilution, deterioration or wrong choice; 2, faulty application, *eg* too little or too much adhesive applied; 3, insufficient control over drying and pressing times; 4, incorrect assessment of the surfaces to be bonded, *eg* using mechanical adhesion when the surfaces require substances employing the specific adhesion principle; 5, use of adhesives having incorrect dry flexibility in functioning and flexing situations, *eg* using brittle glues on book spines that may crack and lift off in use.

MATERIALS

The appearance of books and other finished print may be varied as much by the materials as by the constructional methods used; the cost will also excercise a strong influence upon choice. For these reasons a close study of varieties, standards and usage is desirable.

Paper

Print finishers are expected to process all types of paper and board and an understanding of its structure and furnish will provide a springboard from which to study the manufacturing operations. Some papers may be produced more or less specifically for the finishing operation and have qualities related to the process. Book endpaper material may be of cartridge or similar stock that is relatively tough and dimensionally stable; substance used will depend upon the size and weight of the volume involved but 100 g/m^2 is an average weight. In certain circumstances heavy self-coloured cover-paper stocks, azure laids and good-quality writing papers are used but should have the qualities previously mentioned. Speciality stocks used for endpapers include hand- and mould-made antiques, printed and hand-marbled papers and thin lining-in papers with coloured and patterned surfaces.

Stocks used for strengthening books should themselves be of long-fibred furnish such as kraft and either plain or crêped finish. Speciality papers in this field include brown and white krafts impregnated or coated with latex. These are widely used for lining adhesive-bound or sewn book blocks that are, subsequently, to be rounded and backed.

Packing papers will be chosen from various qualities of pure and imitation kraft for outsides and cheaper casing papers of bulky finish for the inner wraps. Adhesive sealing tapes should be based on good kraft with adequate coating of strong adhesive.

Boards

Three basic types used for bookbinding are strawboard, chipboard and millboard. Strawboard is manufactured from crude straw pulp and has a characteristic straw colour. It is alkaline in nature and by virtue of the stiff straw fibres used a rigid, rather brittle board that tends to give off dust in cutting is produced. The thicker substances are made up of two or more thinner boards pasted together and this can be an occasional source of trouble if not well done. The board surface tends to be rough and shows through if a thin ungrained paper or cloth is glued to it. Interaction between the alkaline board and colour dyes in paper, cloth and ink sometimes leads to leaching of colours and this may be aggravated by the use of alkaline adhesives. The most popular of boards for binding, it is also widely used in rigid box work when it is supplied with one side lined with paper; it is the cheapest of the three types.

Chipboard is manufactured from waste fibrous materials, including paper and board, as a continuous reel. This is subsequently laminated and sheeted to the required caliper and area. Care is taken, during manufacture, to ensure that certain minimum standards of furnish are maintained and the resultant board is softer and more flexible than strawboard. Having much longer fibres it is less brittle and does not dust so easily when cut. The longer fibres make it a suitable medium for the construction of cases and boxes when it can be creased and shaped without fear of fracture. Both strawboard and chipboard are supplied in standard sizes in bundles and with approximate calipers and substances from 0·04mm (250g/m^2) to 4·5mm (3800g/m^2).

Millboard differs from the two previous types mainly in the method of manufacture and density obtained. This board is made of waste fibrous material by building up a continuous thickness of wet fibrous pulp which is later compressed to the required caliper. Considerable curing and stress removal is carried out so that the board may be claimed as dimensionally stable and very suitable for better-quality bookbinding. It is approximately twice the price of straw- and chipboard of comparable caliper.

Large users of boards find it necessary to mature the cheaper varieties as long as possible before use. This is usually completed by unstringing the parcels and stacking them in an environment similar to that in the bindery where they are to be used. Failure to do this may mean that the cut boards will arrive at the casemaking stage in a warped and twisted condition.

Cover materials

Most manufactured cover materials have a basic supporting substance which is impregnated or coated to improve its appearance, handling, wearing or manufacturing qualities and, in the case of cloths, to prevent glue penetration. The most satisfactory of the supporting material is woven textiles, usually cotton, that is produced to various standard weights to suit the end product; a buckram will have a heavier base cloth than, say, a watercloth. These are washed, shrunk and dyed before the coating and finishing can take place.

A starch dough, suitably coloured, is the principal material used to fill the interstices in the cloth. This provides a cheap and satisfactory filling but one that is water soluble; a cloth having a washable and fast-to-light filler is slightly more expensive. Nitrocellulose is used for filling leathercloth, dimensionally stable printing cloths and some waterproof library cloths. These are impervious to water and oil, making the material most suitable for reference books and other situations where a slightly more expensive cloth is acceptable. Polyvinyl chloride is used as a coating for a textile-based material that is similar to leathercloth but is so soft and resilient that it is very difficult to decorate by blocking. This coating is more frequently used in conjunction with a paper base or as unsupported sheet. The latter is extruded in various calipers and qualities for use, in bookbinding, on high-frequency welded work.

Paper-based cover materials are extremely popular because of their cheapness and, as well as the PVC coating, are surfaced with cellulose dope and latex formulations, to give improved qualities to the surface. Another sheet material currently being developed is a polyolefin; this is a tough slightly greasy tinted plastic film, reminiscent of polythene to which it is related.

The habit of producing cloths which have a leather grain is one that dies hard and almost any imitation grain can be imparted to the surface by passing suitably prepared cloth between grained male and female rollers. However, modern preference usually calls for material that shows its natural grain, but where no natural finish exists cloth grains are usually embossed. This is true of the paper felts used as cover materials and these have 'buckram' and 'linen' grains to give improved surface wear qualities. In some instances paper may have a second colour printed upon the surface to give a dual tone effect; in the case of leathercloths and some PVC coated papers, these may be printed, in the grain, to give an antique finish.

Bookcover material is supplied in rolls 1 000 to 1 250mm wide and 45 or 90m in length. All textiles have an unuseable selvedge of 5–10mm width on both edges of the roll and the good surface is usually rolled inside the roll. Paper has a marked machine direction which, of course, runs the length of the roll; the warp direction of textiles also runs the length of the roll.

Leathers

The skins of most animals can be converted to leather and used for book covering, but are normally confined to goat, sheep, pig, cow or calf and occasionally horse.

The resultant leather is described as 'soft tanned' and is a highly flexible product that can be readily moulded. Skins are chosen for size in relation to book areas, *eg* big skins for big books and vice versa. The smallest and probably the most popular skins are those deriving from goats. Two main surface grains are in use: the levant type grain of the Nigerian and similar skins and the well-known hard grain or morocco. These are used for best hand work, library rebinds, bibles, etc.

Full-thickness sheepskin has very little natural grain and is used in this condition for account books and similar commercial applications. This leather can be split into several layers, each being surfaced and prepared identically; these are called skivers, many of which are grained to resemble other leathers for cheap work, *eg* pigskin finish, etc. Roan is a sheepskin with artificial straight grain. When the nap side of leather is prepared to be the appearing surface the prefix 'rough' is used, *eg* rough sheep, and this is also used on a variety of commercial work.

Cowhide has little application in binding, being too thick and stiff to work, but the thinner calf is very popular for the binding of law volumes. Rough calf is also a commercial leather. Pigskin has a distinctive grain, the surface showing the bristle holes in groups of three, most of which can be seen on both sides of the skin. It is a tough material, difficult to pare and consequently reserved for extra large volumes and loose-leaf binders. A thinned or split variety is very popular for high-class diary and fancy goods work.

Vellums and parchments are various skins and animal membranes prepared with the aid of a lime process. The resultant material is a tough film of transluscent character that has a very long life as book covering. Several qualities and types are available; classic or roman vellum being similar in price to good quality nigerian goatskin. Because regular rectangles are being cut from the irregular skin shapes considerable wastage occurs. The total area of the skin is marked on the underside and it is priced according to quality.

Threads and cords

Sewing threads are manufactured from linen, cotton, nylon and terylene. Linen thread is expensive and most suited to handwork; individual thickness is made up of several twists of different calipers. The rule here is the thinnest thread has the highest number, *eg* 22 × 2 cord (two twist of 22 caliper) is thinner than 16 × 2 cord and 16 × 3 cord is thicker still.

Cotton is available for machine work on 'cops', *ie* cone-shaped holders of various weights, and may be bleached, unbleached or coloured to suit the job in hand. Caliper is chosen to suit the machine and the job and as twist may be right hand or left hand care is taken in its choice. Nylon and terylene are not now normally used for booksewing but the latter has been combined with cotton to provide blended threads that give good tensile strength for caliper used yet includes the excellent working qualities of cotton.

Other cords used for stringing and hangers include rayon twists in a variety of calipers, qualities and colours and these may be purchased on cards or reels for machine use.

Miscellaneous textiles

Tape and webbing used for booksewing is usually stiffened for ease of use and is supplied in rolls of various width; for machine work this is important as these have sometimes to fit between needle positions. Hemp is a string-like material made from flax and has very long fibres; different thicknesses or twists are used.

Textiles for strengthening include mull, calico and jaconet, all woven from cotton and suitably starch dressed. Mull has an open weave and quality varies from a type that has a warp and weft about 3mm apart, another having quite a close weave and others with woven 'tapes' or reinforcements.

White calicos are cotton fabrics, starch filled and calendered; qualities vary with use from a very soft type to the pre-shrunk variety used on calico lining machines. Brown jaconet is a much heavier weave material favoured by the commercial binder for strengthening purposes.

Registers, or markers, are the small silk ribbons habitually placed in bibles etc, for place marking. Quality, colour and width are governed by the job. Headbands, when used, are stuck on to imitate the hand-woven versions found on fine bindings. At best these are silk fabrics around a cane centre; at worst a rayon twist stitched to a cotton ribbon. Machine types are available on rolls whilst the best quality is wound around a board to preserve flatness.

Wire and metal units

Wire for stitching and spiral binding is made of low-grade medium carbon steel in metric caliper.

Table 5 Comparative calipers of stitching wire

SWG	*Caliper*	*Metric wire*
	mm	mm
18	1·2192	
20	0·9144	
		0·9
		0·8
22	0·7112	0·7
		0·65
		0·6
24	0·5588	0·55
		0·5
26	0·4572	0·45
28	0·3759	0·4
		0·35
30	0·3048	

Raw metal tends to rust and a coating may be used both to prevent corrosion and to be decorative; these include copper, aluminium and various coloured lacquers and plastic surface treatments.

Eyelets are the small metal units used to prevent a hole in paper or board from tearing out under stress. These are put in by hand or machine; both diameter and depth will be governed by the job and may be brass, steel or aluminium, plain or coloured as required.

Press fasteners are sometimes used as closures on printed products and in connection with high-frequency welded articles. These are composed of two units for both male and female halves and have a special tool, hand or machine mounted, for application; size of stud and depth accommodated can both be varied. A form of loose leaf is constructed by a similar device to the press fastener. It has an elongated male section on to which the punched paper is threaded; closure is by the press stud.

Rivets are used extensively in the fastening of metal parts into paper products. These are usually of the two part tubular type that present a finished head on both sides of the job.

Companies using Wir-O and plastic comb bindings will often purchase the units in diameter, length and colour to suit the job rather than attempt to hold large stocks of miscellaneous sizes. Plastic combs range in diameters from 4·4mm to 32mm, in length up to 42 teeth and a wide variety of colours including gold and silver finish.

Miscellaneous materials

Dyes for ruling and book-edge colouring are usually available in powder form for self-mixing but large users sometimes find liquid colour a more convenient form. Varnish is used to give a finish to some leather volumes and this should be a clear medium-thin type that will not discolour the leather. Polyeurathene varnish is favoured in some binderies. Paper varnish is used for the protection of documents and display notices after suitable preparation with gelatine size.

Powdered or crystal egg albumen is used for glaire when gold leaf is to be tooled on book covers. The powder is allowed to absorb water for 24 hours and then thinned to the required consistency. An alternative used by some finishers and blockers is 'blocking powder'; this is a white powdered shellac that becomes sticky when reactivated by the heated brass tool.

Paraffin wax is sometimes used as a lubricant on paper drills; beeswax is used for imparting a shine to coloured, marbled and gilded book edges. Dilute acids appear occasionally as etch fluids to help the adhesion on difficult surfaces.

The charts in the Appendix show a range of materials with typical characteristics. These can be varied under certain circumstances and if a sufficiently large batch is ordered the customer can have a degree of choice in area, caliper, surface finish, length of roll and other non-basic characteristics. When smaller quantities are ordered these must be selected from the standard range offered by the manufacturer.

PREVENTION OF DETERIORATION

The extent to which a producer of consumer articles can build in protection against future deterioration may be a question of economics but often some protection can be obtained at little or no extra cost. Paper products are frequently transported to parts of the world where environmental conditions are quite different from those in the production centre and this may create problems. Warm moist conditions readily soften some adhesives and exposure to such a situation may cause the closing adhesive of a carton or case to burst open when under stress. Similarly, dry atmospheres tend to reduce the moisture content of books, inducing an unwanted warp to appear in the boards. Prior knowledge of the destination of a product will help to anticipate these situations.

Many books and some technical magazines and papers have to reside in libraries and other storage situations for long periods and are frequently subjected to both biological and insect attack. The environmental conditions in the library are very important and often, even in quite difficult climates, the librarian can take simple steps to prevent the worst of the effects. There are, however, some basic rules that may be applied by the producer to help minimise the deterioration of printed matter, particularly books, which are complained of by overseas customers.

Insects

A wide range of termites, boring insects and cockroaches attack paper products in store. These small boring animals mainly attack the text paper and boards, riddling it with small holes as they move through the stock. Prevention here is primarily the responsibility of the librarian although some papers appear to be less susceptible than others.

Undisturbed books are sometimes chewed by rodents who recognise the animal adhesives as a food source, but by far the most prevalent attack is that made by cockroaches on bookcovers. They graze progressively over the cover picking out the starch from between the interstices of the cloth and leaving a characteristic random pattern on the surface. As the cockroach is primarily concerned with the starch content of the surface it will usually leave cloths coated with non-starch substances undisturbed. These include paper felts, leathercloths, PVA coated papers, nitrocellulose coated cloth, varnished and laminated materials. Any one of these is preferable to starch filled cloth for export to a country likely to be subject to high humidity and cockroach attack and may, perhaps, be incorporated into the binding at little or no extra cost.

Fungi

In atmospheric conditions above 70 per cent relative humidity at temperatures above 22°C moulds and fungi grow readily. The spores are usually airborne and require a receptive surface upon which to grow and mature. Board, paper, cloth and some adhesives are, in various measure, all receptive materials, animal adhesives providing a particularly good nutrient base.

So the problem of how books can be produced for non-temperate climates so that they are least open to attack by mould and insects is one that may have to be faced by the publisher and his binder. The simple solution is to use, wherever possible, the range of cover materials previously listed and to avoid the inclusion of any starch or animal adhesives. Unfortunately the latter point is quite difficult to put into operation as the bulk of binding machinery has been designed around animal glue because of its qualities of economy and fast initial tack. Although some use is now made of PVA (which is almost untouched by mould growth) for first gluing, lining and casing in of publishers' case books, the cases are still made using animal glue.

Various attempts have been made to add substances to the glue which are objectionable or poisonous to the animal that may feed on the book. But to be successful the pesticide used must discourage a wide range of attackers and in any case the pest only dies after feeding on the volume.

Wilfred J. Plumbe in his book *The preservation of books* (Oxford 1964) recommends an addition of 5 per cent endrin (19·5 per cent) to glue, or to treat the glue and starch paste with beta-naphthol.

Appendix

Materials used in print finishing processes

	Name	Raw material	Finish	Characteristics
COVER MATERIALS Textiles	Starch filled	Lightweight cotton	Plain or embossed	Interstices filled with coloured starch dressing
	Patent filled	Lightweight cotton	Plain	Fast to light and washable
	Art canvas Buckram	Heavier cotton weave	Natural canvas Calendered smooth	Very strong Rather stiff
	Leathercloth	Various weights of base cloth	Plain or embossed	Nitro-cellulose filled
	Waterproof library cloth	Light to medium weight base cloth	Smooth and plain	Waterproof and dimensionally stable
	Printing cloth			
Papers	Felts	Chemical wood and cotton fibres	Grained or smooth	Surface fibres protected by light sizing coating etc
	Special felts	Cotton fibres	Imitation leather grains	Simulated leather finish
	Polyvinyl chloride (PVC) coated	Cellulose fibres	Grained or smooth	Coating of various thickness; may be printed

Materials used in print finishing processes

Caliper or width	*Area or length*	*Unit of purchase*	*Use/remarks*
0·965m	45m	Rolls	Medium and cheap work
0·965m	45m and multiples	Rolls	Medium and best-quality books
0·965m	45m and multiples	Rolls	Reference books and strengthening purposes
1·2 to 1·8m approx	45m and multiples	Rolls	Stationery, loose-leaf, library, reference books, etc
0·965m	45m and multiples	Rolls	Library work
			Pre-printed covers
1·0m approx	90 to 100m	Rolls or sheets	Cheap to medium priced publishers' cases, particularly pre-printed work
1·0m approx	90 to 100m	Rolls or sheets	De luxe editions of publishers' bindings
1·0m approx	90 to 100m	Rolls or sheets	Publishers' work, boxes, files, loose leaf, etc, where waterproof and medium-priced material needed

Materials used in print finishing processes

	Name	*Raw material*	*Finish*	*Characteristics*
Polyvinyl chloride	Unsupported PVC sheeting	PVC	Grained or smooth	Coloured, opaque or clear
Leathers	Levant	Goatskin	Open grain	Soft tanned and easily moulded
	Hard grain morocco	Goatskin	Small pimple grain	May be surface dressed
	Basil	Sheepskin	Smooth grain loose back texture	Glazed or unglazed
	Rough sheep	Sheepskin	Nap side dressed	Natural colour
	Skiver	Sheepskin	Skin split 2 to 5 times, little strength	Surface dressed to resemble other grains
	Pigskin	Full thickness pigskin	Distinctive grain shows bristle holes in groups	Hard and tough
		Split pigskin	Shows grain, smooth, dressed	Hard wearing
	Law calf	Calfskin	Putty/light brown in colour	No marked grain
	Coloured calf	Coloured version of above	Smooth and delicate texture	No marked grain
	Rough calf	Thicker calfskin	Nap side dressed	No marked grain
THREADS	Linen	Linen	Polished and/or waxed	Good tensile strength
	Cotton	Cotton	Bleached or natural	Works well on sewing machine
	Blended	Terylene core, cotton outer	Bleached or natural	Very strong for caliper

Materials used in print finishing processes

Caliper or width	*Area or length*	*Unit of purchase*	*Use/remarks*
Calipers used from 0·25 to 0·46 mm	90 to 100m	Rolls or sheets	Exclusively for high-frequency welding process
	Average 0·75m²	Skins	Best letterpress and some library work
	Average 0·75m²	Skins	Mass production work: bibles, prayer books, etc
	Average 1·0m²	Skins	Commercial and stationery work
	Average 1m²	Skins	Stationery work
	Average 1·0m²	Skins	Small books, diaries, boxes, cheap bindings
	Average 1·25m²	Skins	Large volumes, loose leaf, etc
	Average 1·25m²	Skins	Prepared for fancy goods including desk and personal diaries, stationery, etc
	Average 0·85m²	Skins	Binding of law volumes and refurbishing of antique books
	Average 0·85m²	Skins	Letterpress bindings
	Average 1·0m²	Skins	Larger account books and loose-leaf binders
Thin = 22 × 2 cord Medium = 16 × 3 cord Thick = 16 × 4 cord		Per kilo (in hanks)	Hand sewing
Average 0·5mm		Per kilo (on cops)	Machine sewing and stitching
Average 0·25mm		Per kilo (on cops)	Machine sewing

Materials used in print finishing processes

	Name	Raw material	Finish	Characteristics
STRENGTHENING MATERIALS	Tape	Cotton	Stiff or semi-stiff	Various qualities
	Webbing	Cotton	Herringbone weave	Thick; great strength
	Mull (scrim or crash)	Cotton	Open weaves of various qualities	Lightly dressed
	Calico	Cotton	Various quality weaves	White, heavily dressed. May be pre-shrunk for book lining
LININGS	Crêpe kraft	Long-fibred paper	Crêped	Strong and flexible
	Latex impregnated paper	Long-fibred paper	Smooth or crêped	Elastic and expandable
	Case hollow	Short-fibred paper	Rough, bulky, uncalendered	Bulky but firm
DECORATION	Headband	Rayon, cotton or silk	Alternate colour bands. Silk may have cane centre	—
	Marker or register	Cotton, silk or rayon ribbon	Various	—
ENDPAPERS	Cartridge	Esparto and chemical wood stocks	Soft drawing cartridge often used	Rough surface, dimensionally stable
	Self-colour	Various quality cover stocks	Semi-smooth	Single colour; should be fairly stable

Caliper or width	*Area or length*	*Unit of purchase*	*Use/remarks*
5, 7, and 10mm wide	100m	Rolls per 100m	Hand and machine sewing
10 and 15mm	100m	Rolls per 100m	Account-book sewing
1m approx	45m	Rolls	Strengthening book spines, map lining, etc
1m approx	45m	Rolls	General lining and reinforcing
Various	Various	Rolls	Second lining, publishers' edition work
Various	Various	Reels	Lining of, publishers' binding, particularly rounded and backed work
Various	Various	Reels	Case hollows for publishers' books
Depth to suit books with various size squares	—	Rolls for machine	Decoration on better work. Silk/cane centre for best hand work. Cotton and rayon types for longer runs and machine-bound books
Various widths and colours	—	Cards of gross m	Place markers in bibles and other works of reference
110–140g/m^2	Various	Reams	Tipped ends for publishers' work and for made endpapers in hand work. (Other good-quality plain papers may be used, *eg* Azure ledger, etc)
110–140g/m^2	Various	Reams	In place of cartridge or in the make-up of endpapers for library or letterpress styles

	Name	*Raw material*	*Finish*	*Characteristics*
ENDPAPERS *continued*	Marble paper	Printed or hand marbled	Wide variety of patterns and qualities	Available on a variety of different type and weights of stock
FINISHING AND BLOCKING MEDIUMS	Leaf	23 carat gold	Colour according to alloy used	Very thin beaten gold
	Foil	18–23 carat gold	Various colours	On cellulose or polyester carriers
	Foil	Aluminium	Various metallic colours including simulated gold	On polyester carrier; fade and oxide resistant
	Foil	Metallic powder	Imitation gold and silver	On cellulose and glassine carrier; tends to tarnish readily
	Foil	Colour pigments	Glazed or matt effect	Colour may be thick and cause loss of definition
ADHESIVES *Animal*	Scotch glue	Residual animal waste, *eg* bones, skin, etc	In block, bead, powder or liquid form	Tacky, fast drying; brittle film
	Flexible glue	Plasticised animal glue of good quality	Usually block or liquid	Slower tack and drying; readily extendable
Vegetable	Paste	Starch (root, cereal, etc)	Slow moving semi-solid	Near white, high moisture content, slow drying
	Gum	Dextrine or exudation from trees	Thin, near-liquid of low viscosity	Slow initial tack; re-moistening quality is important

Materials used in print finishing processes

Caliper or width	*Area or length*	*Unit of purchase*	*Use/remarks*
Various	Various	Sheet or ream	Account books and various letterpress styles
	83 × 83mm approx	Per 1000 in books of 25 sheets	Hand finishing and edge gilding
Width of roll to suit job	Length of roll 27m approx	Per 25mm on reel	For hand finishing and blocking on all classes of work
Width of roll to suit job	Length of roll 61m approx	Per 25mm on reel	Blocking all styles
Width of roll to suit job	Length of roll 61m approx	Per 25mm on reel	Now seldom used on bookwork
Width of roll to suit job	Length of roll 61m approx	Per 25mm on reel	Blocking all styles, particularly cheaper work and coloured panels
		Kilo	Situations where glue film does not have to flex or fold
		Kilo	Spine of books, etc, where glue has flexing function
		Kilo	Paper to paper, board, leather, etc, where drying can be slow and under pressure; excellent bond
		Kilo	Envelopes, labels, stickers, etc

Materials used in print finishing processes

	Name	Raw material	Finish	Characteristics
Synthetic	Resin emulsion	Polyvinyl acetate (PVA)	White liquid, thin creamy consistency	Initial tack usually poor; film highly flexible and thermoplastic
	Hot melt	Copolymers, resins and waxes	100% solid at room temperature. Working temperature around 160°C	Open time very short and formulated to suit machine
WIRE	Steel	—	Plain	Round or flat
	Steel	—	Anodised or copper coated	Round or flat
	Steel	—	Plastic or cellulose coated	Round or flat
COMBS	Plastic	—	Blocked or plain	Wide or narrow backs as required; variety of colours
BOARD	Straw	Waste straw	Lined or unlined surface	Rigid, cheap, short fibred and brittle
	Chipboard	Wastepaper products including some woodfree pulp	Less rigid and brittle; more expensive	Less rigid, longer fibres giving better bendability
	Container board	Waste cellulose fibres	Kraft lined	Dense grade with specific qualities for its purpose
	White-lined chipboard	Waste cellulose fibres	White lined with various quality outer layers	Fair printing surface, good bending and creasing qualities

Materials used in print finishing processes

Caliper or width	*Area or length*	*Unit of purchase*	*Use/remarks*
		Kilo or litre	Wide range of formulations available for adhesive binding, book lining, laminating, boxwork, etc
		Kilo	Mostly adhesive binding machines and carton sealing
9 diameters between 0·90mm and 0·40mm	Rolls or spools	Kilo	Traditional wire stitching and spiral binding
9 diameters between 0·90mm and 0·40mm	Rolls or spools	Kilo	Situations where rust-resistant qualities are important
9 diameters between 0·90mm and 0·40mm	Rolls or spools	Kilo	Better-quality work in both stitching and spiral binding
Diameter 5 to 28mm	Up to 42 teeth long	Per 100 lengths	Wide variety of letterpress and stationery work
0·5 to 4·7mm	812 × 518mm 762 × 635mm	Kilo	Books, boxes, cartons and containers, showcards, etc
0·5 to 4·7mm	812 × 518mm 762 × 635 mm	Kilo	Books, boxes, cartons and containers, showcards, etc
0·5 to 5·0mm	Makings to suit job	Kilo	Transport containers
300–1 120 micrometres	Makings to suit job	Kilo	Cartons of medium-to-good quality

	Name	*Raw material*	*Finish*	*Characteristics*
BOARD *continued*	Triplex board	Sandwich of bleached and unbleached sulphite	White lined	Good printing, bending and creasing qualities
	Duplex board	Unbleached sulphite lined one side with bleached sulphite	White lined one side rough	Good printing, bending and creasing qualities
	White woodpulp	Wood fibres	May have smooth or rough texture	Clean appearance, fair folding qualities
	Solid white board	Wood fibres	A lined variety of white woodpulp	
	Millboard	Wastepaper products including a % of wood-free pulp	Semi-smooth	Very dense and rigid

Materials used in in print finishing processes

Caliper or width	*Area or length*	*Unit of purchase*	*Use/remarks*
300–1 120 micrometres	Makings to suit job	Kilo	Most widely used carton board. Some qualities of duplex and triplex may be coated for superfine printing
300–1 120 micrometres	Makings to suit job	Kilo	Most widely used carton board. Some qualities of duplex and triplex may be coated for superfine printing
300–1 120 micrometres	Makings to suit job	Kilo	Showcards, calenders, etc
300–1 120 micrometres	Makings to suit job	Kilo	Frozen food packs
1 to 6mm	1016 × 660mm 1016 × 762mm 762 × 635mm	Kilo	Fine leather bindings: library work; best-quality ledgers and loose leaf; boxes and solander cases

Paper sizes and subdivisions in the USA

Paper is manufactured in a wide variety of sizes, calipers and qualities to suit the end products manufactured. Table 1 lists a range of standard paper types and some of the more common sizes available. In each case the area used as a basis for weight (or substance) is indicated, *eg* for bond papers the substance is based on the area 17 × 22 inches. Thus 500 sheets of paper with a basis weight of 24 actually weighs 24 pounds in that size. When the quantities involved are sufficiently large it is common practice for the paper to be specially made and cut into reels or sheets to suit the job.

Subdivision terminology is similar to that discussed in chapter 1. Quarto is a division into four parts, octavo into eight parts; duodecimo into twelve parts; sextodecimo into sixteen parts and so on. Thus a sheet of paper $22\frac{1}{2} \times 17\frac{1}{2}$ inches folded to octavo produces an untrimmed section of $8\frac{3}{4} \times 5\frac{5}{8}$ inches. A frequently quoted list of book sizes is shown in Table 2 with the binding edge shown as the first dimension. It is, however, not always the practice to put the binding edge first and in any order or specification the edge to be bound should be clearly indicated.

Table 1 American book sizes

Name	*Inches*	*Name*	*Inches*
Medium 32mo	$4\frac{3}{4} \times 3$	Medium 12mo	$7\frac{2}{3} \times 5\frac{1}{8}$
Medium 24mo	$5\frac{1}{2} \times 3\frac{5}{8}$	Demy 8vo	$8 \times 5\frac{1}{2}$
Medium 18mo	$6\frac{2}{3} \times 4$	Small 4to	$8\frac{1}{2} \times 7$
Cap 8vo	$7 \times 7\frac{1}{4}$	Broad 4to (up to 13 × 10)	$8\frac{1}{2} \times 7$
12mo	$7\frac{1}{2} \times 4\frac{1}{2}$	Medium 8vo	$9\frac{1}{2} \times 6$
Medium 16mo	$6\frac{3}{4} \times 4\frac{1}{2}$	Royal 8vo	$10 \times 6\frac{1}{2}$
Crown 8vo	$7\frac{1}{2} \times 5$	Super Royal 8vo	$10\frac{1}{2} \times 7$
Post 8vo	$7\frac{1}{2} \times 5\frac{1}{2}$	Imperial 8vo	$11\frac{1}{2} \times 8\frac{1}{4}$

Sizes quoted are not absolute and may vary in practice.

Standardisation

An attempt to standardise basis weights and sizes has been made by the introduction of the MM Paper System. This has been endorsed by the Printing Industries of America and relates substance to 1 000 sheets of 1 000 square inch area.

A United States Congressional report recently submitted that 'if American manufacturers are to escape serious new setbacks in the world marketplace, they must convert to the metric system' and recommended a ten year program of changeover. Most industrial countries are already using or have initiated metric programs and consideration of that part of the ISO system concerned with paper may be justified.

In chapter 1 mention has been made of the A, B and C series of paper sizes and their inter-relationship. It will be noted that publishers in the United Kingdom are unable to go all the way with standardisation and have introduced their own 'metricated book sizes'.

Substance (or grammage) is based on the 'grams per square metre' concept and a single number composed of not more than three figures specifies the substance of the stock regardless of the area of the sheet. This makes possible almost instant comparison of two or more stocks without the need to refer to tables or to make calculations.

Paper sizes and subdivisions in the USA

Table 2 American paper sizes

TYPE OF PAPER	17″ × 22″	17″ × 28″	17½″ × 22½″	18″ × 24″	18″ × 31″	19″ × 24″	19″ × 25″	19″ × 26″	19″ × 28″	20″ × 25″	20″ × 26″	20″ × 28″	20″ × 30″	20½″ × 24¾″	21″ × 32″	22″ × 24″	22″ × 25½″	22″ × 28″	22″ × 29″	22″ × 34″	22½″ × 22½″	22½″ × 24½″	22½″ × 28½″	22½″ × 34½″	22½″ × 35″	22½″ × 42½″
Newsprint															x	x										
Bible																										
Opaque circular	x		x																	x						
Text																										
Book (coated)																									x	
Book (uncoated)			x				x																		x	
Offset book			x																x						x	
Gummed	x									x																
Rotogravure																										
Label											x															
Writing	●	x				x														x						
Manifold	●	x				x									x					x						
Blanks																		●				x	x			x
Bond	●	x	x			x						x					x			x					x	
Ledger	●	x				x														x	x			x		
Blotting						●																				
Box cover											●															
Cover											●															
Manuscript cover					●																					
Glassine																										
Hanging																										
Poster																										
Railroad manilla	●	x			x														x							
Tough check																	●									
Post card																						●				
Tag																							●			
Mill Bristol																							●			
Wedding Bristol																							●			
Index Bristol														x									x		x	
Tissue																										
Waxed paper				x																						
Wrapping tissue				x									x													
Kraft				x									x													
Safety	x	x				x		x	x											x						

Key ● *Basis size*
× *Some other standard sizes*

Table 2 American paper sizes (*continued*)

TYPE OF PAPER	23″ × 29″	23″ × 35″	23″ × 43″	24″ × 36″	24″ × 38″	24½″ × 24½″	25″ × 38″	25″ × 40″	25½″ × 30½″	26″ × 40″	28″ × 34″	28″ × 38″	28″ × 42″	28″ × 44″	30″ × 40″	32″ × 44″	33″ × 43″	34″ × 44″	35″ × 45″	35″ × 46″	36″ × 48″	38″ × 50″	38″ × 52″	40″ × 48″	41″ × 54″	48″ × 60″
Newsprint				●			x				x		x					x			x	x				
Bible							●						x	x		x			x			x				
Opaque circular	x	x					x				x								x			x				
Text	x	x					x			x									x			x				
Book (coated)				x			●			x			x	x		x			x		x	x				
Book (uncoated)	x	x					●						x	x		x			x			x				
Offset book							●						x	x		x			x		x	x	x		x	
Gummed							●																			
Rotogravure							x						x	x		x			x			x				
Label							x			x			x	x		x			x		x	x			x	
Writing					x						x															
Manifold					x						x															
Blanks			x								x			x			x									
Bond	x				x						x							x	x							
Ledger					x	x					x															
Blotting					x																					
Box cover				●																						
Cover	x	x								x										x						
Manuscript cover																										
Glassine				●				x							x											
Hanging				●																						
Poster				●																						
Railroad manilla					x						x							x								
Tough																										
Post card																										
Tag				●											x											
Mill Bristol																										
Wedding Bristol																										
Index Bristol									●										x							
Tissue				●																						
Waxed paper				x																						
Wrapping tissue				●																						
Kraft				●											x									x		x
Safety					x							x														

Key ● *Basis size*
× *Some other standard sizes*

Bibliography

ADAM, A & CO LTD *The Story of Animal Glues*

BFMP *Book Impositions* (1962); *Going Metric with the Printing Industry*; *Metric Sizes for Book Production* (1969); *Production Control for the Printing Industry*; *Safety in Print (Guillotines)* (1968). British Federation of Master Printers

BP *Specification Manual of Printing Machinery and Equipment* (1971). British Printer

CORDEROY, JOHN *Bookbinding for Beginners* (1967). Studio Vista

COOK, J G *Your Guide to Plastics* (1964). Merrow

COUPE, R *Science of Printing Technology* (1966). Cassell

CUNEO CO *Plan for a Good Book* (1957). Cuneo Co, Chicago

DARLEY, LIONEL S *Bookbinding Then and Now* (1959). Faber and Faber

DIEHLE, E *Bookbinding Vol I (Background)*, *Vol II (Technique)* (1946). Rinehart, New York

FAHEY, H & P *Finishing in Hand Bookbinding* (1951). Fahey, San Francisco

FAXON, W O & FOGG, R E *Mechanical Aspects of Adhesive Binding* (1963). T A G A Proceedings

HUTCHINGS, R S & SKILTON, C (Eds.) *Modern Letterpress Printing Vol* 3 (1963). Skilton

JENNETT, S *The Making of Books* (1969). Faber and Faber

MILLS, C J & MILLS, C G *Paper Cutting Machines and Techniques* (1949). C G Mills, Pittsburgh

NAPM *Paper, its Making, Merchanting and Usage* (1956). National Association of Paper Merchants

PAGLIERO, L *Retail Stationers Handbook* (1949). Stationers Association

PATEMAN, F & YOUNG, L *Printing Science* (1971). Cassell

PLUMBE, W J *Preservation of Books* (1964). Oxford University Press

STANLEY, E C *High Frequency Plastic Sheet Welding* (1961). Radyne Ltd

STRAUSS, V *The Printing Industry* (1967). R R Bowker Co

TAPPI *Adhesives*

UPTON, P B & BUSBY, G E *Strength of Unsewn Bindings* (1951). Printing Industry Research Association

VICTORY KIDDER *Guillotine Handbook*. Victory Kidder Ltd

VAUGHAN, A J *Modern Bookbinding* (1960). Skilton

WHETTON, H (Ed.) *Practical Printing and Binding*. Odhams Press

WILLIAMSON, H *Methods of Book Design* (1966). Oxford University Press

Index

Index

Index

Index